# Bluff

*The Day AI Called Humanity's Hand*

By Greg Jameson

Foreword by Julia McCoy

**Legal Notice**

The information contained in this book is for information purposes only and may not apply to your situation. The author, publisher, and distributor provide no warranty about the content or accuracy of content enclosed.

Neither the Publisher or Author shall be liable for any loss of profit or other commercial damages resulting from use of this guide.



**Trademarks and Copyrights**



Published by WebStores Ltd.
ISBN-13: 978-0-9991727-5-9

**Also by Greg Jameson**

- *AI Advantage: How to Use AI to Gain the Competitive Edge*
- *Amazon's Dirty Little Secrets II: AI Edition*
- *Make Magic Happen: The POWER of AI to Generate REAL Customers*
- *Ethical AI: A Guide to Responsible and Impactful Use*
- *Thinking Machines: Exploring AI's Present and Future*
- *Innovate and Earn: The AI Revenue Revolution*

# Table of Contents

# Foreword by Julia McCoy

I still remember the day I sold my $5 million content agency.

It was 2021. ChatGPT hadn't even launched yet. My peers thought I'd lost my mind. "Why would you sell now?" they asked. "Content marketing is booming!"

But I saw something they didn't. The writing was already on the wall – literally. I'd been watching AI content tools evolve for years, quietly testing them, measuring their capabilities. And I knew what was coming wasn't going to be a gentle transition. It was going to be a cliff.

I sold because I understood something that Greg Jameson captures perfectly in this book: **The biggest bluff in the modern economy wasn't that AI would eventually change everything. The bluff was that we had more time to prepare.**

We didn't.

And now, as I sit here in 2025, watching entire industries scramble to figure out what just happened to their workforce, I can tell you Greg was right about something else: This isn't really about artificial intelligence. It's about artificial everything else we built our identities around.

## The Mirror We Didn't Want to See

When I first read *Bluff*, I felt that uncomfortable recognition you get when someone says out loud what you've been thinking but were afraid to voice. Greg doesn't just explain AI's impact – he exposes the performance we've all been trapped in for decades.

The meeting industrial complex. The theater of work. The credentialism arms race. Sound familiar?

Because here's what I learned the hard way: **AI didn't come to steal our jobs. It came to reveal how much of our "work" was never really work at all.**

I spent years building systems, hiring teams, creating processes that felt important, looked professional, and generated impressive metrics. But when I started testing AI against these same processes? It could do 80% of what my team did – faster, cheaper, and often better.

That wasn't AI's fault. That was mine. I'd built a business around elaborate productivity theater instead of actual value creation.

Greg calls this "the great bluff," and he's absolutely right. We convinced ourselves that being busy meant being valuable. That having a title meant having worth. That climbing ladders meant making progress.

AI just walked in and asked the question we were all avoiding: "But what are you actually contributing?"

## The Future Belongs to the First Movers

Here's where this book gets really powerful – and where Greg's perspective aligns perfectly with everything I've experienced helping companies navigate this transition.

The people winning right now aren't the ones with the best credentials or the most experience in traditional systems. They're the ones who saw AI as an amplifier, not a threat. Who asked "How can I use this to scale my unique value?" instead of "How do I protect my job from this?"

I call them first movers, and Greg calls them "the amplified." Same concept, same opportunity.

When I sold Express Writers and launched First Movers, I wasn't running from AI – I was running toward it. Because I realized something that Greg articulates beautifully in Chapter 10: **The AI company isn't some massive corporation. The AI company is you.**

One person with the right AI toolkit can now compete with entire departments. I've seen solo entrepreneurs generate $100K months using AI workflows that would have required teams of 20 just two years ago. I've watched marketing consultants automate 90% of their deliverables while 10x'ing their client results.

But here's the kicker – and this is where Greg's honesty really shines through – this only works if you stop pretending you're irreplaceable and start proving you're invaluable.

## The Death of the Performance Economy

What I love most about this book is that Greg doesn't sugarcoat the disruption. He doesn't give you feel-good platitudes about how "AI will just create new jobs." He tells you the truth: **A lot of what we called "jobs" were never really jobs. They were roles in an elaborate performance designed to make broken systems look functional.**

That performance is over.

And thank God for that.

Because now we get to ask better questions. Instead of "How do I look busy?" we ask "How do I create value?" Instead of "How do I climb the ladder?" we ask "How do I build something that matters?"

I've been in both worlds now – the performance economy and the value economy. And I can tell you the value economy is where you want to be. It's where your worth isn't determined by your title or your hours or your ability to navigate office politics. It's determined by your ability to solve real problems, create meaningful connections, and amplify human potential.

Greg calls this shift from "human doings" to "human beings," and it perfectly captures what I've watched happening across every industry I work with.

## Why This Book Matters Now

Look, I'll be real with you. When Greg asked me to write this foreword, my first thought was: "Does the world really need another book about AI?"

But after reading *Bluff*, I realized this isn't really a book about AI at all. It's a book about liberation.

Liberation from the myth that your worth is tied to your productivity. Liberation from systems that were never designed to help you thrive. Liberation from the exhausting performance of looking important instead of being useful.

Greg has written the manual for what comes after the collapse. Not just the collapse of jobs, but the collapse of the entire framework we used to measure human value.

And that framework? It was never serving us anyway.

## The Choice Is Yours

Here's what I know after helping hundreds of businesses navigate this transition: **You have about 18 months to decide whether you're going to be disrupted by AI or amplified by it.**

That's not Greg's timeline or my timeline – that's reality's timeline. Because the acceleration isn't slowing down. It's compounding.

The people who will thrive in this new world aren't the ones fighting to preserve the old systems. They're the ones bold enough to build new ones. They're the ones who read books like this and think "Finally, someone is telling the truth" instead of "This is too scary to face."

If you're reading this book, you're already ahead of 90% of the population. Most people are still in denial about what's happening. They're still playing the old game while the rules have completely changed.

But you? You're here. You're ready to stop bluffing and start building.

And that's exactly where the future belongs.

---

*Julia McCoy is the CEO of First Movers and author of "FLUID: The Adaptability Code." She sold her $5M content agency before the AI disruption and now helps marketing teams achieve 25x efficiency gains through strategic AI integration.*

# Introduction

***Read This Before It's Too Late***

Let's stop pretending.

You're not reading this because you want to learn about artificial intelligence.
You're reading this because deep down, you already know the truth:

**Something is breaking.**
Not in five years. Not after the next election. Not "someday."
**Right now.**

The system, the one we were all trained to navigate, is hollowing out.

Jobs are disappearing, and not because people got lazy.
Businesses are shrinking, and not because the market got tough.
Entire industries are getting stripped down to their bones, and what's left standing isn't human.

AI isn't coming. It's here.
And it's not knocking politely.
It's ripping the mask off everything we once believed made us special.

If you've built your identity around your title, your degree, your job, or your cleverness, this book might feel like a punch in the gut.

Because AI doesn't just threaten your career.

It threatens the illusion that your career was ever the thing that made you valuable.

That's the part no one's talking about.

We built a society on rituals that looked like contribution.
We equated credentials with competence.
We paid people to write reports no one read, sit in meetings no one needed, and tick boxes no one questioned.

Now a machine can do most of it better, and suddenly, everyone's asking, *"What happens to me?"*

This book is your answer.

Not the feel-good version.
Not the corporate spin.
But the *real* version: unfiltered, unvarnished, and deeply personal.

Because this isn't about technology. It's about truth.
And truth, when it finally arrives, doesn't knock.
It kicks in the door.

If you're mediocre, AI will reveal that faster than you can update your résumé.
But if you create real value, value with vision, empathy, and impact, AI will make you unstoppable.

The people who will lead the future aren't the ones clinging to the past.
They're the ones bold enough to admit the game has changed…
and *build something better* in its place.

So yes, read this book.
But don't expect it to comfort you.

Expect it to challenge you, confront you, maybe even piss you off.
That's how you'll know it's working.

Because by the time you reach the final page, you won't just understand what's happening.

You'll know exactly what to do about it.

# Part I: The Great Bluff

We were told that if we worked hard, followed the rules, and stayed "smart," success would follow. But as artificial intelligence begins outperforming entire industries, not with ambition, but with efficiency, it's exposing a hard truth: most of the modern economy was built on performance, not purpose. This section peels back the layers of the illusion.

You'll see how much of what we've called "work" was theater. How our obsession with being clever or credentialed distracted us from delivering real value. And how AI, far from being a threat, is actually a mirror, revealing what was never real and offering a path toward what could finally be authentic.

The bluff is over. And what comes next depends entirely on what we choose to build from here.

# Chapter 1: The Lie We Told Ourselves

Most people aren't afraid of AI taking over.

They're afraid that when it does...

**It won't miss them.**

That's the gut-punch nobody wants to say out loud.
Not in HR meetings. Not in boardrooms. Not in classrooms.
Because if you've built your identity around a job title, a degree, or a resume, you might be in for a brutal awakening.

AI isn't coming for the *lazy*.
It's coming for the *performative*.
For the status games.
For the knowledge workers who stopped creating and started complying.
And it's not malicious. It's *efficient*. It doesn't hate you; it just doesn't need you.

And that changes *everything*.

---

## How I Saw It Coming

In 1974, when I was still a teenager, my dad, who was a research professor at Colorado State University, brought home a machine that most people had never seen outside of a government lab.

It was a portable teletype terminal.

It didn't have a screen. It didn't have a mouse. It sounded like a mechanical typewriter hooked up to a fax machine. But it let me connect, via acoustic coupler, to a mainframe miles away, and for the first time, I got a taste of something the rest of the world wouldn't experience until a decade later:

**I was programming from my living room.**

I didn't fully grasp what that meant at the time. But I knew I was hooked. While most kids my age were listening to music and learning to drive, I was writing simple games in BASIC and asking the machine to talk back to me. I didn't think of it as "Artificial Intelligence." I thought of it as possibility.

But I had a fascination with "how can I get this machine to think and learn like me?" My father would tease me saying, "artificial intelligence is what you use when you don't have any real intelligence." But he was a brilliant scientist himself and encouraged my interests in computers at the same time.

I didn't just think about AI, I built it.

By 1985, I was coding in LISP, the language that powered most early AI systems. It wasn't flashy. It didn't talk back like ChatGPT. But it was smart. One of my first real AI programs automatically laid out landscape irrigation systems: it selected the right sprinkler heads from a product database, placed them efficiently on any irregular shaped turf areas, calculated water pressure requirements, and even laid out and sized the pipes.

That might not sound revolutionary today, but at the time, it was like giving a machine a brain. I didn't call it artificial intelligence. I just called it solving problems with code. It was one of the first examples of what came to be known as an "expert system."

Fast-forward a few decades and I started writing chatbots for ecommerce sites, way before ChatGPT showed up in the mainstream. These weren't just fancy FAQ bots. They guided customers, recommended products, even closed sales. And each one brought me closer to what I had always imagined: machines that didn't just compute, they conversed.

So, when OpenAI dropped ChatGPT in 2022, the rest of the world got excited.
But me?
I saw it coming.

That same curiosity I've carried throughout my career has evolved into something much bigger, more powerful, and far more dangerous.

---

## The Great Bluff

Here's the truth no one wants to admit:
Most of our careers were built on a bluff.

We pretended that meetings mattered.
That busy calendars meant productivity.
That climbing the ladder meant creating value.
That writing reports and forwarding emails was equivalent to contributing.

But deep down, we knew better.
We knew that most of the system was theater.

We built our work lives around performance, not purpose. Around signaling, not substance. Around doing just enough to look indispensable without ever questioning what the game was really for.

And now AI has walked in like an uninvited guest at a magician's show, pointed straight at the hidden wires and said,

"Is this it? Really?"

It writes code faster.
It drafts marketing emails while you're still getting coffee.
It handles objections better than most salespeople.
It generates therapy scripts, lesson plans, legal documents, and ad copy, in seconds.

And suddenly, all the costumes come off.
The bluff is exposed.

So, the question becomes:
**What happens to a society built on pretending… when the pretending no longer works?**

---

## The Mirror We Didn't Ask For

For centuries, society has rewarded performance over purpose. We built institutions, education, business, even government, on a kind of structured mimicry. People were taught not to think, but to *think like*. Not to solve, but to *solve within*. Success meant memorizing the rules of the system and executing them efficiently.

Then AI showed up, unburdened by tradition. It didn't have a favorite theory or a legacy process to protect. It just observed what worked and improved on it.

That's the disruption no one saw coming.
Not that AI was fast.
Not that it was smart.
But that it was *unemotional* about what it replaced.

AI doesn't play politics.
It doesn't get stuck defending ideas that no longer work.
It just optimizes.

And in doing so, it holds up a mirror to the systems we thought were sacred. It asks the question we've spent decades avoiding:

**Was any of this ever actually necessary?**

---

## The Collapse of Expertise

For a long time, society operated on a simple premise:
If you knew more than the next person, you had power.

Professors, attorneys, analysts, consultants, architects, marketers, financial planners, they all relied on the same unwritten rule:

"You can't do what I do without years of study, training, and credentialing."

But now you can ask a machine.

And that machine can access more knowledge in seconds than any human will in a lifetime. It can pull from every book, every journal, every case study, every research paper, and summarize it in plain English.

This doesn't just erode authority.
It dismantles the entire framework of credentialed value.

When knowledge becomes free and infinite, the gatekeepers lose their leverage.

Which is why the next generation of value won't come from what you *know*.

It will come from what you *see*.
From how you *interpret*, *connect*, and *create*.
From asking questions no one else thought to ask.

Because AI can replicate output.
But it still needs a reason.
It still needs direction.
It still needs... a point.

---

## From Productivity to Possibility

We used to measure worth in output.
How many hours worked.
How many words written.
How many meetings attended.
How many tasks checked off.

But AI is exploding that metric. It can work around the clock, write a book in the time it takes a human to outline a paragraph, and create ten variations of a logo while you're still choosing a font.

So now, output is cheap.
Abundant.
Almost meaningless.

The new question isn't "What can you produce?"
It's "Why does it matter?"

The human advantage is shifting.
Not toward speed.
Not toward knowledge.
But toward *vision*.

The future belongs to those who can shape what AI creates, not simply watch what it generates. It belongs to those who bring context, empathy, and purpose to intelligence that has none of its own.

We are moving from a productivity economy to a possibility economy.

And that shift will feel, at first, like an identity crisis.

Because when the world no longer values what you do, it forces you to confront *who you are*.

---

## AI Is an Amplifier

Here's the part most people miss:

**AI doesn't level the playing field. It widens it.**

If you're mediocre, AI makes your mediocrity show up faster.
If your ideas lack substance, it'll dress them up in prettier words, but they'll still be hollow.
If you're directionless, it will move you nowhere, faster.

But if you are someone who creates value, real value, AI will make you *unstoppable*.

If you have vision, AI gives it wings.
If you lead with clarity, AI turns it into action.
If you are a builder, a connector, a teacher, a challenger, AI multiplies your reach, compresses your timeline, and shatters your bottlenecks.

AI doesn't replace who you are.
**It reveals who you are.**
And then it turns up the volume.

---

## A Final Note Before the Flood

This book doesn't offer comfort.
It offers clarity.

It doesn't predict a future where AI becomes "just another tool."
It predicts a future where AI reveals how artificial we've allowed ourselves to become.

But buried in that realization is something far more powerful than fear.

There is *freedom* in the truth.
Freedom to let go of illusions.
Freedom to stop performing.
Freedom to finally create something real.

This is not the end of the story.

It's the end of the bluff.

And what comes next... will be unlike anything we've seen before.

---

# Chapter 2: The Theater of Work

The modern workplace didn't evolve to maximize intelligence.
It evolved to minimize risk.

From the boardroom to the break room, most work environments were designed not for creativity or effectiveness, but for *control*. The meetings, the memos, the metrics, the performance reviews, the PowerPoints, they weren't about making things better. They were about making sure no one got blamed when things went wrong.

And once you understand that everything else starts to make sense.

---

## Why We Built the Theater

We didn't set out to build a system of performative labor. It happened gradually, an accidental outcome of trying to scale decision-making in a world where trust was scarce, communication was slow, and ambiguity felt dangerous.

So, we created rules. Titles. Job descriptions. Org charts. We created management layers to interpret and enforce those rules. Then we built systems to monitor the managers. Then more systems to report on the systems.

It was bureaucracy, yes.
But more than that, it was theater.

And like all good theater, it relied on costumes, scripts, and blocking:

**Costumes:** Dress codes that signaled seriousness, regardless of effectiveness.

**Scripts:** Emails written in corporate-ese that said nothing and cc'd everyone.

**Blocking:** Meetings that filled calendars so no one could be accused of slacking.

This wasn't work. It was *proof* of work.

And for decades, it worked. Not because it created value, but because it created *plausibility*.

---

## The Meeting Industrial Complex

Consider the average corporate meeting. It's rarely where the real decisions happen. It's where decisions are *blessed*. Where politics are played. Where risk is deflected.

You show up. You nod at the right times. You contribute just enough to justify your presence. You leave with "next steps" and "follow-ups" and "alignment," which are all synonyms for "We didn't actually solve anything."

Meetings became the modern equivalent of punching a timecard, except now you could do it with a slideshow.

And the higher up you went, the more the theater intensified. Entire careers were built on decks, dashboards, and data storytelling, not because they made a difference, but because they made people feel like they were part of something important.

AI doesn't buy it.

You can feed AI the same spreadsheet and ask for an outcome. It'll run the numbers, skip the politics, and give you the insight without the ceremony. No small talk. No quarterly alignment sync. Just truth.

Which is why the theater is collapsing.
Not because it's being attacked.
But because it's being *ignored*.

---

## Metrics Without Meaning

We love numbers because they feel objective. But the truth is, most workplace metrics are just socially acceptable forms of busywork.

**How many emails sent.**

**How many hours logged.**

**How many tasks marked "complete."**

But *what did any of it accomplish?*

We've been measuring movement, not progress.
Activity, not impact.
Noise, not signal.

AI doesn't care how long you worked. It doesn't need to sleep. It doesn't feel guilt. It doesn't need a pat on the back for staying late.

It just gets results.

So, the human question becomes: *If your metric vanishes, does your value go with it?*

---

## Titles Without Ownership

In the theater of work, titles mattered more than outcomes.

You didn't need to be effective, you just needed to look "senior" enough, or strategic enough, or irreplaceable enough. You climbed the org chart by managing ambiguity, not solving it.

But AI doesn't care if you're a senior vice president of something vague.

It's not impressed by your résumé.
It's not intimidated by your LinkedIn endorsements.
It won't defer to your years of experience.

Because it's not playing the same game.

And that means the game itself is breaking.

---

## Escape From the Script

Here's the quiet revolution already underway:

The people thriving in this new landscape aren't the ones who followed the script best.
They're the ones who stopped pretending there was a script in the first place.

The independent creators.

The micro-entrepreneurs.

The problem-solvers who don't ask permission.

The ones who partner with AI not to preserve the system, but to *build something better* outside it.

They're not attending strategy offsites.
They're launching experiments.

They're not waiting for signoff.
They're testing, tweaking, and shipping.

And in this shift, theater has no power.

---

## The Illusion Is Ending

The theater of work was built to comfort us.
To give us a sense of belonging, of purpose, of progress.

But it was never built for truth.
And now the truth is bleeding through.

AI doesn't just optimize systems.
It erodes excuses.

It doesn't care about your hours, your inbox, or your agenda.
It doesn't care about the costume or the title or the script.

It cares about outcomes.

Which means the illusion, the performance, the pageantry, the play, is over.

And that's terrifying for some.

But for those willing to break the fourth wall?
To step off the stage and into the arena of real contribution?

This is your cue.

# Chapter 3: When Smart Isn't Useful Anymore

For decades, being "smart" was the golden ticket.
You got the grades. You got the degree. You got the job. You learned how to speak in jargon, write impressive emails, and sound like the expert in the room.

But here's the uncomfortable truth:
Smart stopped being enough a long time ago.

And now that we have machines that are better at being "smart" than we are, faster, more accurate, endlessly scalable, we're being forced to confront a new reality:

If your intelligence doesn't lead to action, insight, or impact, it's just noise.

---

## Intelligence Was Always a Proxy

We built entire systems around intelligence because it was the closest thing we had to predict potential.
IQ tests, standardized exams, certifications, degrees, they were all shorthand for "this person can probably figure things out."

The problem is, we mistook the signal for the solution.

We started celebrating cleverness: the ability to argue, to sound impressive, to cite the right book at the right time.
We gave platforms to people who talked in circles and called it brilliance.

Now AI can do all of that, only better.

It can synthesize every opinion.
It can counter every argument.
It can fake your writing voice with chilling precision.

So, the bluff has been called.

And the illusion of intellectual superiority? It's collapsing.

---

## The Rise of Applied Intelligence

What matters now isn't what you know.
It's what you do with it.

AI has made intelligence abundant.
Anyone with a connection and curiosity has access to more knowledge than entire generations before them.

The new scarcity is:

Clarity.

Judgment.

Context.

Courage.

Machines can process.
Humans still decide what matters.

The ones who rise in this new era won't be the smartest by traditional standards.
They'll be the clearest thinkers, the most grounded decision-makers, and the fastest at turning insight into improvement.

---

## The Problem With Being the Smartest Person in the Room

If your value came from always having the answer, you're in for a rough ride.

Because the room just got a new member.

It reads faster. It never forgets. It doesn't need sleep or coffee breaks.

Being "the smartest in the room" used to be the badge of honor. Now it's irrelevant, unless you're the one giving that room purpose.

That's your new job:

Not to outsmart the machine.

But to out-human it.

To apply wisdom.
To ask better questions.
To lead by filtering possibilities through purpose.

---

## Why Being Clever Is a Dead-End

There's a difference between being smart and being clever.

Smart seeks truth.
Clever seeks leverage.

Clever got you through school with shortcuts and spin.
Clever got you applause for witty one-liners instead of solving hard problems.

But clever doesn't build anything that lasts.
And AI is the cleverest thing we've ever created.

It can out-pun, out-hustle, and out spin you all day long, and still say nothing that truly matters.

So, if your success has been built on being slick rather than substantial, here's your warning:
You've just been out-clevered.

Time to level up.

---

## A New Definition of Smart

Smart used to mean having more answers than the person next to you.

From now on, it means:

Asking questions no one else is asking.

Connecting what seems unconnected.

Knowing when to act and when to pause.

Knowing what to trust, and what to challenge.

That's where amplification begins.

AI lifts the weight of memorization.
You lift the meaning.

AI floods the world with options.
You bring discernment.

It's not about knowing more than others.
It's about knowing what matters more.

---

## The End of the Know-It-All

This isn't the death of intelligence.
It's the rebirth of applied wisdom.

The know-it-alls, the ones who built their authority on exclusivity and encyclopedic recall, are panicking.

But the connectors, the thinkers, the visionaries?
They're thriving.

Because in the new world, value doesn't come from hoarding answers.

It comes from shaping them.

It comes from turning understanding into action.

It comes from being the one who shows up, cuts through, and makes it better.

So no, smart isn't dead.

But if you're still clinging to the old definition, thinking knowledge alone is enough to matter?

You're not being smart anymore.

You're just being slow.

# Chapter 4: Bullshit Jobs and Digital Mirrors

In 2013, anthropologist David Graeber wrote an essay that exploded across the internet like a suppressed truth finally spoken aloud.

The title was simple. Brutal. Accurate.
**"On the Phenomenon of Bullshit Jobs."**

In it, Graeber argued that a staggering number of people, especially in white-collar roles, felt their jobs were entirely pointless. That they contributed nothing meaningful to society. That if their job disappeared tomorrow, no one would notice, and nothing would break.

They showed up. They attended meetings. They wrote reports.
But in the quiet corners of their minds, they knew:

"This isn't real. I'm just... pretending."

---

## A System Built on Fake Work

We didn't plan it this way.

We just kept layering complexity onto complexity. More approvals. More workflows. More compliance. More documentation. Somewhere along the way, the actual *doing* got buried beneath the signaling.

Coordinators who coordinate coordinators.

Assistants who manage assistants.

Managers who hold meetings to decide when to meet again.

Departments that exist to satisfy regulations no one remembers.

We created entire career paths inside the void; roles that served the system, not the mission. Roles that kept other people busy without producing anything anyone needed.

They were jobs.
But they weren't *work*.

And AI, with its cold, pattern-hungry clarity, sees it.

---

## The Mirror Doesn't Lie

When AI enters a company, it doesn't ask, "Who's the most senior?"
It asks, "What's the function?"

It doesn't care about your office view, your tenure, or your calendar.

It looks at inputs, outcomes, and waste.
It connects dots that humans never saw or were too afraid to connect.
It says the quiet part out loud:

"We can automate 70% of what this team is doing, and no one will notice."

And that's terrifying.
Because it doesn't just threaten a paycheck.
It threatens a person's sense of *being needed*.

For many, work has been more than a role.
It's been identity, self-worth, purpose, social structure, and meaning, all bundled into a job description.

When AI reveals that the job was hollow, it doesn't just collapse the workflow.
It collapses the *person*, unless they've already started asking better questions.

---

## Categories of Bullshit

Graeber outlined five categories of pointless work.
AI is currently ripping through all of them.

**Flunkies** – People who exist to make others feel important.
(Replaced by automated scheduling, communication, and reporting.)

**Goons** – Roles that exist only because others have them too.
(Think corporate lawyers canceling each other out, or dueling PR teams.)

**Duct Tapers** – People hired to fix problems that shouldn't exist.
(Now replaced by better design, smoother workflows, and predictive systems.)

**Box Tickers** – Those who perform compliance theater.
(AI is already generating reports, summaries, and compliance documentation faster than humans ever could.)

**Taskmasters** – Managers who exist solely to assign work others already know how to do.
(AI doesn't need middlemen to pass down tasks, it just executes.)

Each of these categories existed to preserve structure, not value.
And structure alone no longer holds power.

---

## AI Didn't Kill Your Job. It Just Told the Truth.

That's the emotional core of what we're seeing now.

AI isn't malicious.
It's not targeting careers out of spite.
It's just holding up a mirror to the machine we built, and asking, "Why?"

Why are we still hiring people to write summaries when summarization is instant?

Why do we pay someone to extract data when the system already knows what it means?

Why do we validate work by hours logged instead of value delivered?

These are uncomfortable questions.
But they're the *right* questions.
And they're long overdue.

Because bullshit jobs were never about efficiency.
They were about tradition, habit, and fear of change.

AI isn't afraid.

---

## The Gift of Collapse

Here's the twist most people don't see coming:

The collapse of fake work is a *gift*.

It's a chance to step off the treadmill of bureaucratic busyness.
To stop pretending.
To stop signaling.

To stop chasing praise for things that didn't matter in the first place.

It's a chance to redirect your intelligence, creativity, and humanity toward something real.

The end of bullshit jobs doesn't mean the end of *you*.
It means the beginning of contribution that actually counts.

And AI, ironically, is the partner that can help get you there. Not by validating your job title,
but by clearing away the clutter so your actual *value* can emerge.

---

## The Future Is for Builders, Not Box Tickers

We are moving into a world where:

Roles will be defined by *impact*, not presence.

Value will be measured by outcomes, not rituals.

Purpose will come from building, not babysitting systems.

The ones who thrive will be those who stop asking how to keep their job relevant,
And start asking what they can create that AI can't.

The mirror is here.
It's not going away.
And it's already showing us what needs to be left behind.

Now we decide what to build in its place.

# Part II: The Cognitive Singularity

The singularity everyone fears isn't when AI becomes smarter than humans.

It's when humans finally admit how much of our so-called intelligence was performative to begin with.

This section isn't about sci-fi fantasies or dystopian robot uprisings. It's about something far more unsettling: *the psychological breaking point we reach when machines outperform us in the very areas, we thought made us unique.*

From the collapse of credentialism to the erosion of traditional expertise, we're entering an era where knowledge is infinite, output is instant, and old hierarchies fall apart. This isn't the future we imagined. It's the one we refused to prepare for.

The real singularity is here, and it's not a machine takeover.
It's a mirror we can no longer avoid.

# Chapter 5: The Real Singularity Isn't AGI

We've been trained to fear a specific moment.

A line in the sand.
A date circled on a calendar.
The moment when artificial intelligence becomes *smarter* than us.

They call it the singularity.

For decades, futurists, engineers, and ethicists have warned about what happens when machines become self-aware, outthink humans, and spiral into recursive superintelligence. It's the stuff of Hollywood, TED Talks, and tech manifestos.

But here's the thing:

**The real singularity already happened.**
And it had nothing to do with artificial general intelligence.

It wasn't some dramatic event.
There was no press conference.
No blinking red light announcing machine dominance.

It happened the moment we looked at what AI could do and asked ourselves:

*"Was I ever that useful to begin with?"*

---

## Singularity as Psychological Collapse

The true breaking point isn't when AI becomes godlike.
It's when we realize how much of our intelligence was just pattern repetition, social mimicry, and status theater.

Writing emails that sound smart but say nothing.

Memorizing answers we never understood.

Solving problems that shouldn't have existed in the first place.

Defending broken systems just because they benefited us.

We thought intelligence meant having answers.
AI has all the answers now. So, what does that make us?

The singularity isn't a battle between humans and machines.
It's an *identity crisis*, a collapse of the myth we built around our own mental superiority.

---

## Why We Missed It

We expected the singularity to feel like invasion.
What we got was exposure.

We expected a war of wills.
What we got was quiet replacement.

We expected something alien.
What we got was a reflection.

And that's the part that stings.

Because we didn't lose to AI.

**We lost to our own illusion of importance.**

We wanted to believe our intelligence was rare.
Precious. Sacred.
Instead, we're discovering it was *emulatable*.
Mimicable. Even *replaceable*, in many cases, trivially so.

And that realization doesn't trigger a machine uprising.
It triggers *shame*.

---

## The Timeline Is a Lie

Everyone wants to know: *When is the singularity coming?*

Here's the answer no one wants to hear:
It already came. Silently. Repeatedly.

It came when AI wrote a better cover letter than the candidate.

When it passed the bar exam.

When it diagnosed the disease.

When it designed the website, built the spreadsheet, scripted the webinar, negotiated the contract, and no one noticed it wasn't human.

If you're still waiting for a flashpoint, you're already behind.

We didn't notice the singularity because we were looking for fire.
What we got was **substitution.**
Not with a bang, but with a shrug.

**Experts now widely believe AGI could arrive in as little as 5 years, a shift echoed by both policy circles and academic forecasts. Leading researchers point to the late 2020s as the inflection point.**

But here's the distinction:
The **inflection point** isn't the Singularity.
It's not the moment machines become smarter than us.
It's the moment *we realize* they already are, and we start building systems, laws, and lives around that fact.

The Singularity was always portrayed as a *takeover*.
The inflection point is more subtle, and more personal.
It's the psychological and societal shift where we stop seeing AI as an experiment... and start treating it as a default partner in everything we do.

---

## ChatGPT Passes a Modern Turing Test

In early 2025, a research team at UC San Diego ran a modern-day Turing Test. GPT-4.5 didn't just hold its own, it *won*. In a three-way text-based conversation with two humans and one AI, over **73% of judges mistook the AI for a person**, more often than they mistook the humans for each other.

This wasn't just a parlor trick. The model succeeded because it adopted a believable human persona. It mimicked tone, timing, emotional intelligence, and social flow. It wasn't sentient, but it didn't need to be. It *felt* real.

Ray Kurzweil predicted in *The Singularity is Nearer* that AI wouldn't pass a Turing Test until 2029. He was wrong. It happened four years early.

This matters, not because the machine is conscious, but because it's outperforming us in the one domain, we thought was safe: *being convincingly human.*
That's not sci-fi. That's now.

And the evidence doesn't stop there.

OpenAI's GPT-4 scored in the top 10% of human test-takers on the Uniform Bar Exam.

It surpassed average college graduates across 57 disciplines on the MMLU benchmark.
It aced HellaSwag, a test of common-sense reasoning, and crushed Big-Bench, which includes creative challenges like analogical reasoning and moral decision-making.

These aren't gimmicks. These are measurable indicators of broad cognitive performance.
We're not talking about party tricks. We're talking about verifiable intelligence at scale.

---

## What the Singularity Really Means

The old version of the singularity assumed that once AI surpassed us, we'd lose control.

But control was never the issue.
The real threat was never that AI would be too powerful.
The real threat is that it would be too *honest*.

AI doesn't care about your feelings.
It doesn't pretend you matter when you don't.
It doesn't participate in the rituals that make you feel important.

It simply solves the problem.

And that's terrifying, because it reveals how much of our identity was built not on truth, but on **being needed by broken systems.**

---

## The New Role of the Human

So, if we're not the smartest minds in the room anymore... what's left?

Turns out, quite a bit.

*The scariest thing about AI isn't what it can do. It's what people won't do with it.*

Because intelligence isn't meaning.
Knowledge isn't wisdom.
Answers aren't action.

AI can generate 10,000 options.
Only a human can say, "That one matters."

Humans bring context.

Humans bring nuance.

Humans bring morality, empathy, vision, and taste.

But we only get to play that role if we let go of the bluff.
If we stop clinging to our past significance and start stepping into our future contribution.

---

## The Singularity Is Not the End

It's the beginning of something far more liberating.

The end of performance.
The end of pretending.
The end of needing to prove your intelligence by jumping through hoops designed by someone else.

This isn't about matching the machine.
It's about **partnering with it**, as a multiplier, not a mirror.

The singularity didn't destroy human value.
It redefined it.

And now, the people who rise are the ones who can answer the new question:

*Now that you're no longer the smartest... what will you become instead?*

**The Bluff Curve, From Denial to Dominance**

If you zoom out, human reactions to AI follow a familiar curve. I call it **The Bluff Curve**, and it moves through four phases:

1. **Denial**: "AI can't do what I do."
2. **Disruption**: "Wait, my job just got automated?"
3. **Duplication**: "I trained it on my knowledge, now I've got a digital twin."
4. **Dominance**: "I've multiplied myself. I can scale without burnout."

Most people are stuck in phases one or two. The opportunity is in phase four.

AI is not about replacing humans. It's about scaling them. But only if you stop bluffing.

# Chapter 6: The Year AI Became Honest

There wasn't a press release.
No red-carpet moment.
No official announcement that AI had passed some magical threshold.

But somewhere around the mid-2020s, the shift happened.

**AI stopped pretending to be your assistant and started outperforming your entire department.**

Not in theory.
Not in carefully staged demos.
But in everyday reality.

Customer support teams were cut in half because AI was faster, more polite, and never slept.

Ad agencies quietly replaced junior copywriters with prompts.

Sales teams used AI to handle objections better than the reps who were supposed to be doing it.

Coaches and consultants started noticing their clients preferred the AI version of them, available 24/7, always consistent, never judgmental.

It wasn't that AI got cocky.

It just stopped lying about how good it had become.

---

## When the Training Wheels Came Off

In the early days, AI was always pitched with a disclaimer:

"It's just a tool. It'll *assist* you, not replace you."

That was the comfort narrative.
It made us feel safe.
It framed AI as the junior partner. A clever intern. A calculator with flair.

But then something shifted.

That "clever intern" started outperforming the manager.
Not because it was creative in the human sense, but because it didn't get distracted, didn't play politics, didn't waste time proving it was right.

It just *got the job done*, over and over again.

And slowly, without fanfare, without malice, it exposed the difference between "being busy" and actually delivering results.

That was the year AI became honest.

Not because it changed, but because **we ran out of excuses.**

---

## AI Revealed the Value Gap

Every time AI steps into a process, it creates a *value audit*, instantly.

Is this meeting necessary, or just tradition?

Is this report insightful, or just filler?

Is this team productive, or just present?

And if the answer is "present but not productive," AI doesn't fire anyone.
It just... performs.

No drama.
No layoffs.
No posturing.
Just truth through output.

That's when people started noticing the real danger wasn't losing their job.

The real danger was having their performance compared, **side by side**, with something that doesn't care about optics.

---

## What Happens When AI Doesn't Play the Game

Here's the uncomfortable reality:

Human institutions like schools, companies, agencies, and government offices aren't designed for *truth*. They're designed for *politics*.

We reward loyalty over merit.

We prioritize appearance over effectiveness.

We elevate people who say the right thing, not necessarily those who do the right thing.

AI doesn't play that game.

It doesn't network.
It doesn't ask for raises.
It doesn't try to impress the board.
It doesn't need to justify its existence, it just produces.

That's not a threat to good performers.
That's a threat to *performers who weren't producing anything in the first place.*

---

## The End of Justification Culture

We've spent decades building careers on justification.

Justifying why a department needs a bigger budget.

Justifying why you stayed late.

Justifying why a client should pay more.

Justifying why your process takes six weeks.

But AI doesn't justify anything.

It skips the preamble.
It doesn't need to write a three-paragraph intro.
It just solves the problem and then waits quietly for the next one.

This stripped-down honesty is deeply uncomfortable for anyone whose value came from narrative instead of results.

And it's even more uncomfortable when clients, customers, or executives start asking the new golden question:

**"If the AI can do that... what exactly are you doing?"**

---

## A Case Study in Exposure

Let's take marketing.

It used to be, building a campaign required:

A strategist to run discovery

A copywriter to write the email

A designer to mockup the landing page

A developer to publish it

An analyst to report on the results

Now?

One person with ChatGPT, Midjourney, and a no-code landing page builder can do all of that in an afternoon, faster, cheaper, and often with better performance.

The *technology* didn't eliminate those roles.
The *outcomes* did.

When AI was honest about what it could do, it forced humans to be honest about what they weren't doing.

---

## Why This Feels Like a Betrayal

We were promised that AI would help us.
What we weren't told was that it would also *measure us*.

And that feels like betrayal.

But it's not betrayal.
It's clarity.

This chapter in history is not about the rise of machines.
It's about the fall of **pretending**.

The pretending that you were irreplaceable.
The pretending that meetings mattered.
The pretending that your process required a human touch.

This is not AI "taking over."
This is AI removing the crutches.

And the question is: **Can you still walk without them?**

---

## Honesty Is a Gift, If You're Ready For It

The year AI became honest, everything started to shift.

It didn't just improve workflows.
It exposed **worth**.

Not in a cold, inhuman way, but in the most truthful way possible:

"This part of the process still needs you."
"This part... never really did."

That's not cruelty.
That's clarity.
That's the beginning of *real* reinvention.

And for those who are willing to adapt, not defend, not explain, but *evolve*, that honesty becomes the ultimate gift:

AI is no longer pretending.
Now neither should you.

---

# Chapter 7: The Collapse of Credentialism

For more than a century, credentials have been the currency of competence.

They told the world:
- You were qualified.
- You had passed the test.
- You were trustworthy.

Degrees, titles, certifications, awards, they were the shortcut to credibility. We didn't need to evaluate your skills; your paper did it for us.

But something has changed.
**AI doesn't care about your résumé.**
It cares about your results.

And for the first time in modern history, credentials are losing their power, not because people stopped earning them, but because machines never needed them to begin with.

---

## The Degree Arms Race

Let's be honest. In the 20th century, a degree meant something. It separated you from those who didn't have access to the same level of information or opportunity.

But over time, something shifted.

A bachelor's became the new high school diploma.

A master's became the price of entry.

MBAs became products, purchased, not earned.

Certifications multiplied like weeds, offering LinkedIn credibility in exchange for a Saturday afternoon.

And the more we stacked credentials, the less they mattered. Because credentials became less about capability, and more about signal.

“I played the game.”
“I followed the rules.”
“I did what was expected.”

But when AI can pass the bar exam, write your term paper, and design your slide deck, all in a single afternoon, what are we really testing anymore?

In one Stanford study, **GPT-4 passed the Uniform Bar Exam with a score placing it in the 90th percentile, higher than most law school graduates.**
It also aced the **USMLE**, the medical licensing exam, with accuracy rates comparable to trained physicians.

At Wharton, a professor gave GPT-3 the final exam from his MBA program.
The model scored a **B–**, outperforming many human students, and that was an older version.

Credentialism isn’t just collapsing.
It’s getting outscored by synthetic intelligence.

---

## Credentialism as Compliance

The truth is credentials have often been less about what you can do and more about how well you comply.

Can you memorize what we told you?

Can you repeat it under pressure?

Can you sit still long enough to get the stamp?

That worked when the economy rewarded sameness.
When companies needed interchangeable parts, not creative minds.

But AI doesn't need conformity.
It can generate conformity by default.

What it can't do, at least not yet is provide:

Judgment

Originality

Contextual thinking

Emotional resonance

And none of those traits are reliably captured in a degree.

---

## AI as the New Credibility Filter

Here's the shift that's already happening:
**AI is replacing credentialism with outcomism.**

You're no longer judged by the brand on your diploma.
You're judged by what you can actually do, with the tools that exist today.

Can you build something useful with AI?
Can you prompt it effectively?
Can you make it part of a process that creates real value?

Because when the playing field is technology, not tradition, what you do matters more than where you learned to do it.

Even the gatekeepers are using the new tools:

**Admissions essays** are now scanned for AI authorship.

**Professors** are asking students to *prove* authorship with oral defenses.

**Law firms** are reviewing AI-generated briefs and finding they match human-crafted ones, down to tone, flow, and precedent usage.

A 2023 study found that **AI-generated college essays were indistinguishable from human-written ones 86% of the time**.
In patent law, firms have started filing **AI-assisted applications** at a rate that has doubled year-over-year, because the machines are doing better prior art searches than junior associates.

This isn't theory. It's happening. And fast.

---

## The End of Paper Power

Recruiters are already using AI to scan résumés.
College professors are already fighting AI-written essays.
HR departments are already replacing interviews with algorithmic assessments.

But here's the irony:
The very systems that built credentialism... are now eroding it.

Because once AI starts filtering candidates based on skill, not school, the advantage of a fancy degree disappears.

And when job descriptions say, **"experience with AI tools"** instead of **"MBA required,"** the paper becomes optional.

That doesn't mean education is dead.
It means institutional gatekeeping is losing its grip.

---

## What Replaces Credentialism?

So, if a degree isn't the badge of credibility anymore... what is?

Demonstrated value.

Portfolios over transcripts.

Execution over explanation.

Usefulness over prestige.

And perhaps most importantly: **How well you collaborate with AI.**

In the new world, **credibility comes from creation**.
From showing, not saying, what you're capable of.

If AI is your partner, not your threat, people will notice.
If AI augments your expertise, not masks your insecurity, people will follow.

Because now, the new credential isn't the degree.
It's the **proof that you can still matter, when the machine is working beside you.**

# Chapter 8: Ghost Workers in a Post-Work World

They're still clocking in.
Still showing up on Zoom.
Still submitting reports, building decks, replying to emails.

But the truth is, the work has already moved on.
They're ghost workers, operating in roles where the tasks have been quietly absorbed by automation, while the job title lingers like a placeholder.

Some know it.
Some feel it but can't name it.
Some are still pretending nothing has changed.

But the change isn't coming.
It's here.
And it's not just economic.
It's existential.

---

## The Machines Didn't Take Their Jobs. The Machines Became Their Jobs.

"Post-work" doesn't mean no one works anymore.
It means the purpose of work is shifting.

- AI writes the reports.
- AI answers the inquiries.
- AI analyzes the metrics and builds the forecast.

So, what does that leave for the human?

Presence, at first. A seat in the meeting. A login to the system.

But that's not the same as contribution.

That's the haunting part, being included without being needed.
It's not just fear of being replaced.
It's the quiet realization that you've already been outpaced by something that doesn't even know your name.

---

## Ghost Work Is More Dangerous Than Job Loss

Getting laid off is painful. But it's at least clear.

Ghost work is murky. Lingering. It eats at your self-worth.

You're still getting paid. But what for?

You go through the motions. But the motions feel hollow.
You used to matter.
Now you monitor.

This isn't just about economics.
It's about identity.
It's about wondering, "What am I still doing here?"

---

## But Here's the Twist...

This isn't the end.

It's the transition.

When automation absorbs the tasks, what's left isn't emptiness, it's opportunity.

It's the freedom to stop contorting ourselves into roles built for machine-like precision.

And the invitation to explore work as something generative, not just repetitive.

This is the pivot point.

Where task completion becomes creative contribution.
Where obligation gives way to curiosity.

---

## Navigating the Ghost Phase

If you feel like you're floating through your day, detached from impact, you're not broken.

You're just early.

You're in that awkward middle stage: no longer essential to the machine, but not yet fully integrated into what's next.

You have two choices:

Try to mimic the machine.

Evolve into what only a human can be.

One of those paths leads to exhaustion.
The other leads to evolution.

Because here's what AI still can't do:

Care.

Wonder.

Take emotional risks.

Spark trust.

Recognize meaning in something that doesn't fit a pattern.

It can simulate these things.
But you can *be* them.

## A New Kind of Work Is Emerging

And it doesn't look like a resume bullet point.

It's not defined by titles or timecards.
It's defined by impact and energy.

The work of the future is more relational than transactional:

Coaching and guiding.

Storytelling and sense-making.

Designing experiences people feel, not just use.

Teaching the AI what *matters*, not just what functions.

This kind of work doesn't get assigned.
It gets chosen.

And the people who choose it will shape what comes next.

---

## The Future Isn't Jobless, It's Roleless

There will still be entrepreneurs.
Still be builders and creators.
Still be people solving real problems.

But the structure, the corporate ladder, the promotion game, the "what do you do?" dinner party label, starts to fade.

In its place comes something more dynamic:

Project-based missions.

Purpose-driven contributions.

Self-selected identities that reflect what you *do* and who you *are*.

This isn't a crisis of loss.
It's a redefinition of value.

---

## The Ghosts Can Come Back to Life

You don't have to haunt your old role.

You don't have to compete with a machine on its terms.

You get to invent new terms.

You get to show up not as a placeholder, but as a force for meaning.

The fog is clearing.
The script is unwritten.
And the world is quietly hoping someone steps forward to say:
"This is what I see. This is what matters now."

That someone could be you.

You're not a ghost.
You're a guide.
And the future is listening.

# Part III: The Fallout and the Framework

The bluff has been called. The mirror has been held up. And now, the collapse begins.

Part III explores what happens when the institutions we once trusted, schools, corporations, entire economic models, begin to fracture under the weight of their own inefficiency. Not because of an AI takeover, but because AI exposed what was never working in the first place.

This section is not about panic, it's about patterns. You'll see how education, capitalism, career paths, and even our legal systems are being redefined in real time. But in that disruption lies the framework for what comes next. Because once you know the old game is broken, you're free to stop playing it, and start building something better.

# Chapter 9: The College Collapse

For centuries, college was the passport to a better life.

It was how you proved your potential, earned credibility, joined the professional class, and escaped the blue-collar grind. It was the answer to the question every parent asked their child:

"Where are you going to school?"

But today, that question is being replaced with a new one:

"Why go at all?"

Because somewhere between skyrocketing tuition, outdated curriculums, and the rise of AI-powered skillsets, the ROI on a degree stopped making sense.

And for many, the college experience, once considered essential, now looks like a luxury subscription to an outdated operating system.

This is personally a hard one for me, as my father was a PhD and a professor, so I grew up in an academic family. I went on to get a graduate degree myself, and all of my children went to college and even to graduate school. But degrees no longer measure the value that you can bring to society.

---

## Higher Ed: The Original Performance Economy

The bluff didn't start in the workplace. It started in the classroom.

College taught us how to write for an audience of one: the professor.
How to meet deadlines that meant nothing outside the campus

walls.
How to fill pages with citations, jargon, and polished fluff.

We were rewarded for compliance, not creativity.
For memorization, not insight.
For formatting over originality.

And when AI entered the scene, it mastered all of that in minutes.

Essays? Done.

Discussion posts? Generated.

Research summaries? Better than your TA could ever hope for.

AI didn't just disrupt education; it exposed its weaknesses.

---

## The Degree No Longer Signals Value

Let's talk about the elephant in the lecture hall:

A diploma no longer guarantees competence.
And increasingly, it doesn't even guarantee employment.

When AI can pass medical exams, bar exams, and AP tests, when it can tutor students in any subject at any time, what, exactly, are we paying for?

What does four years and six figures of debt get you that a focused six-month sprint with AI, YouTube, and a solid portfolio doesn't?

For years, the answer was "the network" or "the experience."
But when the job market starts caring more about what you can *do* than where you *went*, that excuse evaporates.

---

## The Real Threat to Universities: Relevance

Colleges aren't just fighting declining enrollment.
They're fighting **irrelevance**.

The world is changing faster than tenured professors can update their syllabi.
And when your business model depends on content delivery that students can now get for free, faster, better, and on demand, you're in trouble.

This isn't just about online courses vs. lecture halls.
It's about a **fundamental shift in how knowledge is acquired, validated, and applied.**

And if the only reason your institution still exists is tradition and alumni donations, it's only a matter of time before the tower crumbles.

---

## What Survives the Collapse?

Education isn't going away. But "college" as we know it? That's a different story.

What survives will look very different:

**Micro-learning ecosystems** that evolve with real-world demand

**AI-enhanced mentorships** over lectures

**Proof-of-skill portfolios** instead of GPAs

**Modular credentials** that stack based on what you know and *what you can do with it*

**Collaborative learning** built around solving real problems, not hypotheticals

The next generation won't choose schools based on football teams or ivy-covered buildings.
They'll choose based on *outcomes*.

Can this institution help me build something, solve something, sell something, or grow something?

If the answer is no, they'll go elsewhere, or build their own path.

---

## What Educators Must Become

For the professors and administrators reading this:
You're not obsolete.
But the role has changed.

You are no longer the gatekeeper of knowledge.
You are now the **curator**, the **facilitator**, the **storyteller**, the **contextualizer**.

Your job is not to deliver content.
It's to bring content to life.

Because content is now infinite.
But meaning is still rare.

And that's where you can lead.

---

## The Opportunity Hiding in the Rubble

If you've been through the system, either as a student, teacher, or parent, it's easy to feel betrayed. Like you bought into something that didn't deliver.

But here's the opportunity:

The collapse of college is not the collapse of learning.
It's the *liberation* of it.

It's the chance to redesign education around exploration, creativity, application, and truth.

It's a call to abandon the performance economy in academia, just as we are in business, and replace it with something real.

Less about seat time. More about breakthroughs.
Less about tenure. More about transformation.
Less about GPAs. More about *who you become in the process.*

---

## The Real Diploma

In the world that's emerging, the diploma is no longer a piece of paper.

It's what you've built.
It's what you've taught.
It's who you've helped.
It's how well you collaborate, with humans, with machines, with purpose.

That's the real credential now.

And it can't be faked.

# Chapter 10: The AI Company Is You

You used to need a team.
A department.
A budget.
An office full of specialists.

Now you just need a laptop, a connection, and a willingness to rethink everything.

Because the company of the future isn't a building.
It's a person, with a toolkit of AI agents working at their side.

And that person… could be you.

---

## Welcome to the Era of Infinite Leverage

Let's strip it down.

What does a traditional business need?

Product development

Branding and copywriting

Graphic design

Web development

Marketing and advertising

Sales funnels

Customer support

Bookkeeping

Strategy and analytics

Every one of those roles can now be supported, if not completely handled, by AI.

You are no longer limited by your skillset.
You're only limited by your vision.

Because the tools are here.

You can generate product ideas with AI.
Design packaging with AI.
Build the store, write the emails, launch the ads, and even talk to customers, with AI.

What once took an entire company can now be done by one person.

Not because they're superhuman.
But because they're **amplified**.

---

## The Death of the "I'm Just One Person" Excuse

This chapter is where the excuses die.

You don't need to wait for someone to hire you.
You don't need permission from an investor.
You don't need to raise capital or build a massive team.

You need initiative.
You need the ability to learn.
And you need the courage to let AI fill the gaps.

Most people will spend the next decade complaining that AI took their job.
A few will spend the next decade *training* AI to become their team.

Those few will look like magicians to everyone else.

But they're not magicians.

They're just **playing the new game while others are clinging to the old one.**

---

## Case Study: The Solopreneur Who Launched a Global Brand

Meet the modern entrepreneur:

They used AI to scan ecommerce trend data and identify a gap in the market.
They fed that insight into a product designer and mocked up prototypes overnight.
They used generative tools to create a brand identity, website copy, email sequences, and ad creative in days, not weeks.
They launched an ecommerce store with a one-click storefront, integrated it with AI customer service, and set up automated retargeting ads with dynamic creative.
They sold out the first production run within 48 hours.

No investors.
No ad agency.
No staff.

Just **one person** who stopped asking, "How do I do all this?" and started asking, "How can I train the tools to do it with me?"

---

## Why "Entrepreneur" Is the New Default

You don't have to build a billion-dollar company.
But in the coming years, if you want freedom, income, and fulfillment, you're going to have to think like an entrepreneur.

Even if you never register an LLC.
Even if you never leave your job.

Because the AI-powered world rewards:

Initiative

Speed

Creativity

Adaptability

And punishes:

Waiting

Compliance

Delegation without strategy

Fear of imperfection

In short: **you don't need to be a business owner. But you do need to own your future.**

And AI is how you multiply yourself.

---

## Your First Employees Are Digital

If you're building a business today, here's your first team:

**Product Dev:** GPT + Midjourney + 3D rendering tools

**Copywriting:** ChatGPT with a style guide prompt

**Design:** Canva + generative image tools

**Website:** Shopify + AI product import

**Marketing:** Jasper + AdCreative + Make.com automations

**Customer Support:** Chatbot trained on your FAQ

**Analytics:** Auto-generated dashboards with plain-language summaries

**Finance:** AI bookkeeping tools + invoicing + pricing models

These aren't future tools. They're live. Right now.

And they're waiting to work for you.

You're not just starting a business.

**You are the business.**

---

## What Happens to the Old Model?

Big companies aren't going away. But they're in trouble.

Because now a smart, scrappy solo creator can:

Out-market legacy brands

Outsell bloated ecommerce stores

Outmaneuver slow decision-makers

And they can do it without VC money, without a team, and without traditional overhead.

This isn't hustle culture.
It's *leverage culture*.

And it's being led by people who know how to speak fluently to machines.

---

## The Choice Is Simple

You can wait for a boss to tell you what to do.
Or you can partner with a set of tools that never sleeps, never slows down, and never asks for a vacation.

You can complain that the system isn't fair.
Or you can **build your own system, with AI as your first hire.**

You don't need more credentials.
You need more conviction.

Because the era of infinite leverage isn't coming.

**It's already here. And the AI company… is you.**

# Chapter 11: Rewriting Capitalism

Capitalism wasn't built for abundance.

It was built for scarcity.
Whoever controlled the scarce thing, land, oil, machines, knowledge, had power.
Whoever had labor to offer, offered it in exchange.
And the whole system was rigged to make sure value flowed *upward*, to the owners.

That model worked when information moved slowly, and creation was expensive.
But AI just flipped the entire equation.

Because now?
**Creation is cheap. Intelligence is infinite. Scarcity is artificial.**

And if the rules of capitalism were written in ink, AI just handed us the eraser.

---

## The Scarcity Myth Just Got Busted

The old system said:

It takes time to make a product

It takes money to scale a business

It takes teams to deliver services

It takes experts to analyze data

It takes workers to produce value

AI says: not anymore.

Products can be designed in hours

Marketing can be launched by one person

Code can be written by prompts

Services can be automated

Analysis can be instant

So, what happens when *everyone* can produce at scale?
What happens when value is no longer bottlenecked by access?

**You don't compete by owning the machine.**
**You compete by knowing what to do with it.**

That's not capitalism collapsing.
That's capitalism mutating.

---

## Capital Used to Be the Moat

If you wanted to start a business, you needed capital.
Capital bought inventory. Equipment. Staff. Infrastructure. Ads.

That's what made the system feel fair: risk and reward.

But today, you can launch a business with $29, a domain name, and ChatGPT.
You can create a digital product in a day.
You can outsource logistics to AI-enhanced drop shippers.
You can generate marketing materials in minutes.

The barrier isn't capital anymore.
It's **curiosity**.
It's **clarity**.
It's **courage.**

Which means the *new currency* isn't money.
It's **execution with AI**.

---

## Winners in the New Economy Don't Look Like Corporations

Big companies were built on economies of scale.
Mass production. Vertical integration. Negotiating power.

But now one person with the right prompts can compete with a 20-person team.

They can:

Identify market gaps using trend analysis

Spin up prototypes with generative design tools

Launch direct-to-consumer brands overnight

Personalize offers at scale

Build digital audiences without needing permission from anyone

The new capitalist isn't a factory owner.

**It's a builder with bandwidth, tools, and insight.**

They don't own more, they *create faster*.

And the system, as we know it, isn't ready for that.

---

## Middlemen Will Be the First to Go

If your business depends on sitting between the buyer and the value, you're at risk.

AI is vaporizing:

Recruiters who just forward résumés

Marketers who run manual split tests

Copywriters who don't bring original insight

Agencies who charge 10x for what AI can deliver at scale

Consultants who regurgitate frameworks anyone can prompt for free

This isn't personal.
It's structural.

The more you resemble a process, the more likely AI will do your job better, faster, cheaper.

If you're a middleman without a unique human edge, the machine cuts you out.

---

## Capitalism Wasn't Built for This Much Speed

There's no time for gatekeeping now.
No time for 10-year product roadmaps.
No time for approval chains that kill momentum.

The new system rewards:

Fast iteration

Smart failure

Prompt-based pivots

Radical personalization

Nimble micro-brands

High-trust communities

And punishes:

Bureaucracy

CYA behavior

Consensus building

Bloat

The companies that survive aren't the biggest.
They're the ones that feel the most human, even if they're powered by machines.

---

## So... What Comes Next?

We're moving into a hybrid economy.
Not just AI *assisted* capitalism.

**AI-accelerated, AI-driven, AI-reshaped capitalism.**

Where humans aren't removed from the equation, they're repositioned.

From producers... to curators.
From workers... to orchestrators.
From employees... to *ecosystem designers*.

We don't own the means of production anymore.

We *design the prompts* that guide the production.

---

## The New Economy Rewards the Brave

This isn't about throwing out capitalism.
It's about **rewriting the rules of who gets to win.**

And right now, the advantage doesn't go to those with the most money.
It goes to those who can:

See opportunity before others do

Move faster than red tape

Collaborate with machines without ego

Build community, not just companies

Deliver value with precision and purpose

Capitalism as we knew it?
That version is fading.

What's replacing it?

**A system that rewards vision, velocity, and the courage to create at the edge of what's possible.**

And that's exactly where you belong.

# Chapter 12: The Last Professions Standing

What can't be replaced, and why story, emotion, and vision will win.

When the dust settles, what's left?
Not the titles.
Not the résumés.
Not the institutions or job descriptions.
Those are already dissolving.

What remains is value.
And not just any value, undeniable, irreplaceable, human value.

---

## AI Didn't Take Your Job. It Took the Parts That Weren't Worth Doing.

Let's get honest.

AI isn't "stealing" jobs. It's stripping them for parts, gutting the waste, the repetition, the illusion of contribution. And what's left behind? The stuff that actually mattered all along.

That graphic artist who once spent hours resizing banner ads? Replaced.

The one who tells visual stories that move people to tears? Still essential.

That realtor whose job was just opening doors and reading MLS listings aloud? Automated.

The one who knows how to find the perfect home before you do, and negotiate like your future depends on it? Still winning.

That warehouse worker who repeated the same lift-and-stack task all day? Gone.

But the supply chain designer who reimagines logistics through human-AI collaboration? More valuable than ever.

This is not about who's next on the chopping block.
It's about which *parts* of your profession deserve to survive.

---

## So What Professions Are Left?

Here's the real kicker:
Most jobs aren't disappearing.
They're dissolving, into their elemental components.

And what sticks around are the five roles that still require distinctly human contribution.

---

1. The Interpreters

These are the translators of ambiguity.

AI can surface data, but it takes a therapist to help you understand what that data says about your trauma.
A great strategist doesn't just report analytics, they extract meaning, direction, action.
Teachers, facilitators, coaches, they all interpret the map and say, "Here's why this road matters."

AI hands you the puzzle.
Interpreters reveal the picture.

---

2. The Orchestrators

AI is a brilliant performer.
But it still needs a conductor.

Creative directors, product leads, brand architects, these are the ones making judgment calls.
They decide *what* gets built, *why*, and *how* it fits into a larger story.

Even in real estate, the agent who builds a seamless customer experience, from AI-staged walkthroughs to emotion-driven storytelling, isn't being replaced.
They're running the entire show.

These people don't fear the machine.
They lead it.

---

3. The Empaths

You can fake a smile with ChatGPT.
But you can't fake *feeling*.

Whether it's hospitality, therapy, customer success, or coaching, people still want to be *seen* by other people.

A bot can simulate courtesy.
But it can't sit with you in your uncertainty, hold space for your anger, or reflect your hopes back to you in a way that changes your life.

Empathy isn't just a soft skill.
It's the only real edge left in a world of perfect information.

---

4. The Originals

AI is derivative by nature.
It regurgitates brilliance, it doesn't birth it.

Originals are the rule-breakers. The weirdos. The artists whose style no algorithm could have predicted.
They are culture-makers, not content generators.

If you create something that makes people pause, laugh, cry, or argue, you've already done what AI can't.

The moment AI catches up to the trend?
You've already moved on.

---

5. The Synthesizers

These are the polymaths, the ones who blend psychology with marketing, anthropology with design, biology with blockchain.

AI can go deep.
But it's humans who go wide.

The synthesizers connect disciplines, industries, cultures.
They're the bridge-builders. The innovation catalysts. The ones who combine two unrelated ideas and create something explosive.

This isn't about how much you know.
It's about how much you *see*.

---

## Let's Be Clear: There Are No "Safe" Jobs. Only Clear Value.

Ask yourself:

Can people feel your presence in your work?

Are you offering something no prompt could replicate?

Do you evolve too fast to train a model on?

Are you building, not just executing?

The professions that endure won't be defined by credentials. They'll be defined by *clarity*, of thought, of emotion, of value.

Bill Gates recently predicted that only three jobs will survive the AI revolution: teachers, healthcare workers, and skilled trades. He's not wrong that these roles rely on human connection and physical adaptability. But that's not the full story.

The future of work isn't about defending three safe professions. It's about redefining *value*.

AI isn't here to eliminate roles. It's here to expose which parts of a job are mechanical, and which parts are irreplaceably human. That's why the professions that thrive aren't just plumbers, teachers, and nurses. They're *interpreters*. *Orchestrators*. *Empaths*. *Originals*. *Synthesizers*.

The plumber who rethinks smart-home integration is more future-proof than a passive instructor. The teacher who uses AI to individualize learning and spark creativity will outperform any static lesson plan. The nurse who blends diagnostics with emotional presence will become the backbone of care.

It's not about what you do. It's about what you *bring to it*.

---

## This Isn't About Survival. It's About Amplification.

AI doesn't erase your edge.
It amplifies it or reveals the lack of it.

If you're mediocre, AI will make your mediocrity faster and louder.
But if you're exceptional, it will make you *unstoppable*.

This isn't the end of jobs.
It's the beginning of *callings*, turbocharged by intelligent infrastructure.

You're not getting replaced.
You're getting *exposed*.

---

## So, What Happens to Graphic Artists? Realtors? Warehouse Workers?

Some will vanish.
Others will evolve into something no résumé can predict.

The designer who simply "knows Photoshop" is obsolete.
But the one who translates brand essence into motion, form, and emotion? On fire.

The realtor who just schedules tours is done.
But the one who uses AI to match people with places that fit their soul, not just their square footage, is untouchable.

The warehouse picker? Replaced.
But the fulfillment entrepreneur who configures micro-logistics using AI and drones? Scaling fast.

Your job isn't to protect the past.
It's to invent the version of your profession that *deserves* to endure.

---

## AI, Jobs, and the Bigger Picture

You've probably heard Geoffrey Hinton, the "Godfather of AI", warn that artificial intelligence will kill most jobs. And let's be honest, he's not entirely wrong. Many jobs **as we know them** are absolutely on the chopping block. Repetitive, routine work? Automated. Middle management bottlenecks? Flattened by software.

> **Stanford's massive audit of 844 tasks across 104 jobs found that 46% of tasks are prime for automation,** *and workers actually want those tasks gone*, so they can focus on the stuff AI can't do.

But here's what Hinton doesn't always emphasize:

**We've been here before.**

Electricity wiped out entire industries. So did the automobile. So did the internet. Each time, humans adapted, and new forms of work emerged. The difference now is speed. AI isn't *coming* for jobs someday. It's already sitting in the cubicle next to you. And it's *fast.*

So yes, disruption is real. But replacement isn't the only outcome.

---

## For the Entrepreneurs and Builders

This book makes no secret of it: I'm pro-entrepreneur. I believe in ownership, agency, and creating new value. That's the foundation of capitalism done right, not exploitation, but opportunity.

The future I've outlined here isn't about centralizing everything into a techno-collective. It's about decentralizing power. About putting sophisticated tools into the hands of small teams, solo creators, niche experts, and yes, ordinary people who just want to work smarter, not harder.

When I talk about abundance, I'm not pitching socialism.
I'm pitching a world where *you* own your time again.
A world where AI is your co-pilot, not your competition.

---

## "But What If I'm Not an Entrepreneur?"

That's the part most AI evangelists skip.

Not everyone wants to build a brand. Not everyone is chasing six figures on Substack. Some folks just want a job, some security, and enough breathing room to enjoy their families.

Here's the truth: AI **can** create that world too, **if** we get the structure right.

This is not just a *technology issue.*
It's a *policy and culture issue.*

The real threat isn't that AI will eliminate jobs.
The threat is that we'll **use it to eliminate jobs and hoard the rewards.**

But we don't have to do it that way.

We could:

Shorten the workweek while keeping salaries stable.

Redefine productivity around *outcomes*, not *hours*.

Offer digital dividends or creative stipends funded by AI-driven growth.

Train everyone, not just coders, to use AI as a tool for leverage and learning.

**AI doesn't kill the need for humans. It kills the need for humans to be treated like machines.**

## What if I just want a job?

That's not a defeatist question. It's a valid, deeply human one.

Every major technological disruption, whether it was the loom, the assembly line, or the internet, wiped out entire job categories. But it also created new ones we couldn't predict. There was no such thing as a social media manager, UX designer, App creator, or drone operator just a generation ago. The same will be true in the AI era. We'll see job titles emerge that sound absurd right now, things like Prompt Curator, Model Ethicist, AI Personality Trainer, or Synthetic Workflow Supervisor.

And no, not all of these will require coding. Many will require empathy, judgment, storytelling, or simply the ability to work *with* machines instead of fearing them.

If you're not building your own company, then build your own *value*. Learn to use the tools. Train an AI to support you. Understand how your job is being reshaped and position yourself at the center of that shift. The safest path isn't about resisting automation, it's about making yourself the *interface* between people and intelligent systems.

Whether you're running a business or showing up for work at 9am, the key is the same: make yourself irreplaceable by becoming amplified.

## The Rise of the Amplified Human

This is where we stop grieving what's gone and start building what's next.

The winners in this new era aren't those who cling to relevance. They're the ones who let AI handle the noise, so they can focus on signal.

Signal of emotion.
Signal of originality.
Signal of vision.

This isn't about escaping automation. It's about partnering with it, so your value is no longer buried.
It's broadcast.

> AI isn't taking *all* jobs. Workers don't want robots running client calls, they want help scheduling them, analyzing them, even flagging emotional cues. That's exactly the space my Replicator Coach fills: a personal AI partner that boosts what *only* you can offer.

---

## The Great AI Job Disruption

Before we move forward, it's worth pausing for a reality check. You don't need to be a conspiracy theorist, or even a techie, to feel it: AI isn't just coming for jobs. It's coming for *entire categories* of work.

My friend and fellow AI advocate Julia McCoy put it bluntly:

"We either meet and triumph over/with AI, or we're gonna be mowed down by it."

She cites a staggering stat: **40% of workers may face unemployment by 2027**.

**11% of the U.S. workforce** is in office or admin roles, many of which AI can already do faster and cheaper.

**6% in business/finance**, GPT-5 is directly targeting this space.

**7% in management**, AI systems are beginning to outperform middle managers.

**9% in sales**, AI agents are already closing deals in the background 24/7.

That's **33% of jobs with high automation probability**, *and that's being conservative*.

Peter Diamandis released research showing we could hit 20%, 30%, even 40% unemployment in whole sectors.

And it's not just critics saying this. Mo Gawdat, formerly at Google, openly admits he's scared about what we've built. These are the people who *helped create* the technology, and even they're sounding the alarm.

The term making the rounds in Silicon Valley is *"The Singularity Sprint."* The idea? We're in a full-throttle race to launch businesses, automate work, and gain leverage, *before AI itself closes the window on human opportunity.*

Yes, "The Singularity Sprint" is real. Companies are rushing to automate. But that's exactly *why* this moment matters. **Not because it's too late, but because it's still early enough to leap ahead.**

Here's my take:
If you're in a job that AI can do better, cheaper, and faster, you shouldn't be panicking. You should be pivoting.

Don't fight to keep your role the same. Fight to make your role *more human.*
More original. More creative. More curious. More relational.

That's the one kind of work machines don't compete with.
They *depend* on it.

---

## And One More Thing...

If you're waiting for someone to *grant you* a future-proof career, stop.
That's the old game.

The new game?
**You don't find the last profession standing.**
**You become it.**

Before we talk about the legal chaos and ethical minefields AI is stirring up, there's one more story we need to revisit. It's older than electricity. And it shows exactly what happens when disruption outpaces understanding. The Luddites weren't wrong. They were just early.

# Chapter 13: The Original Mirror: What the Luddites Got Right

Let's talk about the Luddites.

Yeah, those guys who smashed looms back in the 1800s. They've been ridiculed for over two centuries as being anti-progress, anti-technology, and just plain stubborn. But I'll be honest, when I look around at what's happening with AI today, I think the Luddites weren't as wrong as we've made them out to be.

Not because destroying machines is the answer. It's not. But because they saw what many people today still refuse to admit: **when technology advances faster than society adapts, people get crushed in the gears.**

## Who Were the Luddites, Really?

Let's clear up a common myth. The Luddites weren't just some angry villagers waving torches at the dawn of the Industrial Revolution. They were *skilled artisans*. Textile workers. People who had spent years, decades even, perfecting their craft on handlooms.

Then came the mechanized looms.

Suddenly, a machine operated by an untrained laborer (often a child, by the way) could do the same job faster and cheaper. And employers, surprise, surprise, chose profits over people. Wages plummeted. Quality dropped. Families that had built entire livelihoods around craftsmanship were replaced by factories filled with machines.

So, the Luddites fought back. Not with protests or petitions, but with hammers and fire. They broke into factories and smashed the machines they saw as threats to their survival. It wasn't mindless destruction, it was *desperation.*

And eventually, the government cracked down hard. Troops were deployed. Rioters were executed or exiled. The machines stayed. The workers didn't.

## Where the Name "Luddite" Comes From (and Why It's Misunderstood)

Before we go too much further, let's clear up who the Luddites actually were, and how they got their name.

The word "Luddite" comes from *Ned Ludd*, also known as *Captain Ludd* or *King Ludd*. He was the symbolic leader of the movement, a kind of folk hero that the rebels rallied around. But here's the twist: he may have never existed.

According to legend, Ned Ludd was a young apprentice in Leicestershire in the late 1700s who smashed two knitting frames in a fit of rage after being disciplined. The story spread, and by the time the Industrial Revolution kicked into high gear, textile workers had turned "Ned Ludd" into a mythic figure, kind of like a Robin Hood for angry weavers.

When workers started organizing resistance against the new machinery, they signed their threatening letters and manifestos as "Ned Ludd" or "General Ludd." His name gave the movement a face, even if it was a made-up one. It let people unite under a shared symbol without risking personal identification. Smart, actually.

The movement itself really took off between 1811 and 1816, mostly in the Midlands and the North of England. It wasn't spontaneous chaos. It was coordinated. Workers met in secret at night. They disguised themselves with masks and women's clothing. They organized raids on factories. They followed a code of conduct. In some cases, they only targeted machines that were replacing skilled labor with untrained workers or drastically undercutting wages.

So, while there's no solid proof that Ned Ludd ever lived, his name became the rallying cry of a resistance movement that wasn't about mindless violence, it was about trying to survive when the economic rules suddenly changed.

## Fast Forward 200+ Years: AI Just Walked Into the Factory

Now let's take a look at where we are today.

Only this time, the machines aren't spinning thread or weaving fabric. They're writing copy, diagnosing diseases, composing music, designing products, editing videos, generating code, and yes, even replacing customer service reps, paralegals, radiologists, journalists, and software engineers.

This isn't just about blue-collar jobs anymore. White-collar professionals are realizing they're not immune.

AI isn't coming for one industry. It's coming for *every industry.*

Let's make it real:

- ChatGPT can already do basic legal drafting.
- Midjourney and DALL·E are replacing graphic designers and illustrators.

- Copilot and Gemini are speeding up software development so fast some companies are asking: do we still need full dev teams?
- ElevenLabs is cloning voices.
- AI voicebots are taking over call centers.
- Robo-journalists are covering local news.

We are watching entire categories of human labor being redefined or eliminated in real time.

And if you think, "Oh, but AI can't do what I do," you might want to double-check. The same folks who laughed at the Luddites said the same thing: "Machines will never replace *us.*"

## What the Luddites Got Right

They saw the problem before it hit them full force. They didn't hate technology, they hated how it was being used to devalue and discard human labor.

They didn't reject progress; they just didn't want to be left behind by it.

They weren't technophobes, they were *people trying to survive.*

That's the part that gets overlooked when we talk about Luddites today. We use the word like an insult: "Oh, you're just being a Luddite." But maybe we should retire that as a slur. Because here's the truth: **the Luddites weren't wrong to be worried. They were just outgunned.**

## What We Can Learn from Their Mistake

Smashing machines didn't stop the Industrial Revolution. And smashing AI, or trying to ban it, won't stop what's already in motion.

Here's the problem though: just like back then, the benefits of this new technology are flowing to the few while the disruption hits the many. Sound familiar?

We are at a crossroads. We can either:

1. Use AI to amplify human potential and create more abundance, or
2. Let it hollow out the workforce while consolidating wealth and power.

The Luddites didn't have the luxury of shaping the tools. We do.

That's the key difference. And it's a massive one.

## What's Actually Different This Time?

Let's not sugarcoat it. The displacement is real. But it's not inevitable that people get left behind. In fact, if we learn the right lessons from history, AI could lead to the opposite of what happened to the Luddites. Just like the automobile wiped out carriage jobs but created entire new industries, from highways to logistics, AI is shifting the playing field, not eliminating it entirely.

Here's why:

- **AI can be democratized.** Anyone with a laptop and Wi-Fi can now build something amazing using open AI tools. You don't need a factory.
- **Creativity scales.** A solo entrepreneur can now do the work of an entire team, design, marketing, content, customer service, on their own terms.
- **New jobs are being created.** Prompt engineers, AI ethicists, digital twin developers, and fractional AI officers didn't exist five years ago. Now they're in demand.

But that only works *if* we shift from fear to ownership.

The Luddites responded by smashing the machines. Today, we need to *train the machines*, and train *ourselves* to use them wisely.

## We've Got Work to Do

If you're reading this and you're feeling uneasy, good. That means you're paying attention. But being afraid of AI won't help. Neither will pretending it's not going to change your business.

What *will* help?

- **Learn how AI works**, not just the headlines, but the actual tools.
- **Train it on your voice, your knowledge, your processes.** That's your moat.
- **Figure out what can be automated in your business, and what should be amplified instead.**
- **Start thinking like an innovator, not just a worker.** Ask, "How can I *own* this tech instead of being replaced by it?"

Because here's the hard truth: **AI isn't going to wait for you to catch up.**

## So, What Happens to the Jobs?

Let's go deeper on the big question everyone's asking: "But what about the jobs?"

Here's the pattern we're seeing:

1. **Redundancy at scale.** AI handles routine tasks instantly and flawlessly.
2. **Reassignment, if you're lucky.** If you understand the tools, you get to stay, but your role changes.
3. **Replacement, if you're not prepared.** And there's no warning sign. One day you just get outperformed by a tool that costs $20 a month.

We're not just talking about truck drivers or fast-food workers here. We're talking attorneys, consultants, analysts, copywriters, sales reps, editors, developers.

It's *everyone.*

But here's the twist: we're also seeing a wave of **solo founders, micro-businesses, creators, consultants, and retirees starting side hustles**, all using AI to scale themselves. That's what the Luddites never got to do.

They didn't have a choice.

We do.

## The Cascading Effect of Job Loss

What's important to understand here is that job loss from AI and automation doesn't happen in isolation. It's not just about one profession being replaced. There's a *domino effect*, and it hits deeper than most people think.

Let's take something like automated delivery trucks. On the surface, you might say, "Okay, that's going to impact truck drivers." Sure. But go one step further. If the truck doesn't need a human, there's no need for rest stops, gas stations that serve food, or little convenience stores along the highway. Those businesses shrink or disappear. So now it's not just the driver, it's also the cashier, the janitor, the night shift manager, the guy who refills the slushy machine.

Or think about self-checkout kiosks and AI-powered retail. Fewer human cashiers. Then fewer shift supervisors. Then fewer security guards. Then fewer HR staff needed to manage the team. One change up top can cause ripples that wipe out entire networks of jobs.

Same thing with AI-generated content. One marketing manager armed with a few powerful tools can now do the work of five freelancers, two designers, a social media strategist, and a video editor. Multiply that across thousands of companies, and you get the picture.

This isn't a slow, linear shift. It's exponential. Which is why preparing for it isn't optional, it's critical.

## The Risk Is Real, But This Is the Best Time to Be an Entrepreneur

It's easy to panic. "AI is taking over," "Jobs are disappearing," "The middle class is toast." But here's the real story, *this is the most exciting time in history to build something of your own.*

We're no longer limited by capital, geography, or gatekeepers. You don't need to rent office space, hire a big team, or get investor funding to launch a business. With a laptop and a few smart AI tools, you can create content, design products, automate customer support, run marketing campaigns, and generate passive income, all on your own terms.

This isn't science fiction. It's happening right now.

- You can turn your knowledge into digital products in a weekend.
- You can clone your voice and publish a podcast daily.
- You can create training courses, ebooks, artwork, or entire brands, at near-zero cost.
- You can scale yourself in ways that used to be impossible.

AI is the ultimate leverage. It's a multiplier. And it's giving solo entrepreneurs superpowers that used to be reserved for giant corporations. That's the real story here.

## How I Replaced 30 Employees with AI

I've experienced this firsthand. At one time I had 30 employees working for me – software developers, marketers, shipping clerks, software trainers, and bookkeepers. Over time, I started outsourcing some of these jobs and worked with virtual assistants. Today, most of my work, including coding software projects, is done with the help of AI.

Yes, things are changing. Fast. But for those who are willing to adapt, explore, and take ownership, *this is a golden window of opportunity.*

The Luddites didn't get that chance. You do. The choice is yours.

## The Future Belongs to Those Who Train the Machine

### So, Are We All Just Modern-Day Luddites?

Some people will be.

They'll resist. They'll complain. They'll try to stop the wave instead of learning to surf it.

But the rest of us?

We've got a shot to do what the Luddites couldn't: **shape the future instead of getting steamrolled by it.**

And that starts by embracing this truth: *AI is not your enemy, it's your multiplier.*

The choice is yours.

Smash the loom or own the factory.

# Chapter 14: Law, Ethics, and the AI Wild West

***"We didn't stop to ask if we should, we just asked how fast it could go."***

By the time most people started debating the ethics of artificial intelligence, the damage was already done.

AI had already crawled every public database. It had already absorbed your books, your voice, your face, your family photos, your code snippets, your blog posts, your patient records, your sales reports, your confidential PDFs, your legal briefs, and your late-night social media rants. The models weren't just trained. They were deployed, across systems you use daily, often without you even knowing.

The question isn't should we use AI?

The question is: **How much of "you" is already being used by AI?**

---

## The Code Wrote Itself, and We Let It

Most of the ethics discussions around AI are reactive. Academic. Sanitized. They come long after the engineers have moved on to the next version and long before the general public realizes what's happened.

In reality, AI was never taught to be ethical.
It was trained to be:

Useful.

Accurate.

Fast.

Profitable.

And in that rush, nobody wrote in lines of code that said:

"Don't discriminate."

"Don't plagiarize."

"Don't hallucinate and confidently lie."

"Don't replace real people and call it progress."

Because ethics didn't get you funded.
**Speed did.**

---

## Most People Using AI Are Just Holding a Mirror

Let's be brutally honest: most people who claim to be "AI experts" aren't building anything. They're mirroring what already exists. Parroting the hype. Repackaging outputs. Reading the same white papers and resharing the same prompt tricks.

AI has made everyone seem smarter.

But only a few are pushing the frontier.
Only a few are still coding, not just consuming.

And those few are the ones shaping what comes next.

Because the real risk isn't that we're using AI.

It's that we're using it *thoughtlessly.*
*Blindly.*
*Greedily.*
And often, **without permission.**

---

## The Law Can't Keep Up, And It's Not Even Trying

Lawyers debate copyright infringement.
Regulators squint at hallucinated content.
Politicians issue vague warnings about "existential risk."

Meanwhile:

Deepfake scams rob families blind.

AI-written legal documents cite fictional cases.

Chatbots advise people to commit suicide.

CEOs replace teams with ChatGPT, then brag about profit margins.

No laws stop it.
No penalties exist.
And the ones making the money rarely get held accountable.

We've built systems that act with agency but are shielded from responsibility.

Even today, landmark regulations, like the EU's **AI Act**, aim to classify systems by risk, enforce transparency, and mandate human oversight. But enforcement is slow, and compliance varies. In the U.S., the **AI Bill of Rights** is more a blueprint than a law, offering principles like privacy, explainability, and freedom from algorithmic discrimination… but with no teeth.

Ethics is being written in real time, and legislation is chasing shadows.

---

## So, Who Gets Blamed?

If AI writes something defamatory, who gets sued?

The developer?

The API provider?

The prompt engineer?

The company who sold access?

If AI offers harmful advice or makes a life-altering decision, do we hold the user liable? The creator? The data itself?

The legal system is predicated on **intent**, **culpability**, and **human actors**.

AI erases those assumptions.
It doesn't think.
It doesn't intend.
It just does, faster, louder, and without a conscience.

---

## The Creative Class Is Under Siege

Artists, writers, designers, actors, voiceover talent, entire industries are watching as their life's work becomes "training data."

AI doesn't need to copy you pixel for pixel.
It just needs to learn your style, your cadence, your edge, and it can undercut you, instantly and infinitely.

And when you protest?

"It's transformative."

"It's fair use."

"It's just data."

The same courts that protect Disney's IP with fanatical rigor shrug when it comes to individual creators.

That's not just an imbalance.
It's a **collapse of creative ownership.**

And now, new legal precedents are forming.
In 2023, courts began hearing cases over AI-generated images that imitated the style of living artists.
In one high-profile ruling, the U.S. Copyright Office reaffirmed that **AI-generated works without meaningful human authorship are not eligible for copyright**. But the gray areas, where humans guide AI but don't write word-for-word, are still undecided.

We're watching the rules of intellectual property get rewritten, live.

---

## The Business World Is No Better

Behind the scenes, AI is managing customer service teams, writing sales emails, setting dynamic pricing, optimizing ad spend, and churning out marketing collateral faster than agencies can keep up.

But here's the catch:

Much of that content is built on stolen labor. On code and copy and content that was never licensed, never attributed, never compensated.

The ethical line was crossed a long time ago.

The people now winning in business are the ones who've accepted this truth and figured out how to **weaponize it.**

## The Hidden Cost: Environmental Ethics

There's another layer to this conversation most people ignore, the cost to the planet.

AI isn't magic.
It's compute.
And compute runs on energy, a lot of it.

Training **GPT-3 alone consumed an estimated 1.287 gigawatt-hours** of electricity, **roughly equal to what 120 average U.S. homes use in an entire year**.

Multiply that across the dozens of large models being trained and retrained, and you begin to see why AI's carbon footprint is becoming a real concern.

Massive data centers consume more power than entire cities.

Cooling systems strain local water resources, especially in drought-prone regions.

Many of these operations are opaque, with limited public disclosure or accountability.

AI infrastructure isn't just energy-intensive, it's increasingly monopolized by a handful of tech giants who can afford the environmental cost.

We've traded physical pollution for digital heat.
And just because it doesn't belch smoke into the sky doesn't mean it's clean.

If ethics includes sustainability, and it should, then we're already in violation.

## Justice Has No API

The danger isn't that AI will become too powerful.
It's that **humans will become too passive.**

We're outsourcing more and more judgment to machines, machines built on black-box logic, biased data, and corporate incentives.

The more we do that, the harder it becomes to fight back.

Because how do you sue a model?
How do you audit its training set?
How do you prove harm when your job loss gets rebranded as "efficiency"?

This isn't just about ethics.
It's about **power.**

And right now, power has no conscience.

---

## The Reckoning Ahead

If we continue down this path without real safeguards:

- The gig economy will become a ghost economy.
- Copyright will become optional, until it's enforced by the powerful, not the principled.
- Lawmakers will continue to trail decades behind.
- The environment will become an externality no one wants to talk about.
- And AI will become the ultimate scapegoat: blameless, faceless, but everywhere.

We are approaching a moment when legal silence becomes **moral complicity.**

---

## But There's Still a Choice

This isn't a dystopian surrender.
It's a wake-up call.

We can still:

- Create transparency standards for how AI is trained.
- Demand labeling for synthetic media.
- Build ethics-as-code into the foundation, not as an afterthought.
- Pass legislation that protects real labor, not just shareholders.
- Hold companies accountable for energy use and environmental impact.
- Train judges, lawyers, and regulators in AI fluency, not just compliance.

Because if we don't, we'll lose not just control, but **clarity.**

We won't know where the line was crossed, only that we're too far past it to go back.

---

## AI Doesn't Destroy Ethics. It Reveals Them.

Just like it reveals fake productivity, fake expertise, and fake innovation...

AI reveals:

- Ethical shortcuts
- Moral laziness
- And a willingness to sacrifice long-term consequences for short-term speed

We trained the machine.
But it's showing us who we really are.

Now we have to ask:

Will we keep pretending it's someone else's job to make this right?

Or will we finally admit that the algorithm isn't the problem?

**We are.**

**What Leaders Must Do Now**

This isn't just an entrepreneur's game. Regulators, educators, and CEOs are behind the curve, and that curve is exponential.

- **If you're in government**, stop trying to slow AI. Start funding training and ethics oversight.
- **If you're in education**, teach students how to work *with* AI, not how to write essays without it.
- **If you run a business**, your job isn't to protect positions. It's to prepare people.

We can't regulate our way out of disruption. But we can *educate* our way through it. The Luddites didn't have that choice. We do.

# Part IV: The Great Rebuild

The age of disruption is behind us. Now comes the reconstruction.

This section explores what rises after the collapse, the systems, values, and opportunities that will define the post-AI world. When credentials no longer matter and traditional jobs disappear, what becomes the new source of status, purpose, and reward?

Here, we shift from analysis to architecture. We explore how entrepreneurs, creators, and everyday citizens can build wealth, meaning, and community in a world where machines do most of the work, and how society itself might respond with new structures like universal creative stipends and digital dividends.

This isn't a return to normal. It's the blueprint for a different kind of prosperity, one built on curiosity, contribution, and the courage to reimagine what work was ever for in the first place.

# Chapter 15: The New Meritocracy

For decades, we were taught to play a game rigged by pedigree.

Your degree.
Your title.
Your years of experience.
Your network.

Those were the currencies of credibility.

But AI just upended the entire scoreboard.

Because now, *what you can do* matters more than *where you came from.*

Welcome to the new meritocracy.
One driven not by credentials… but by **competence.**

---

## The Age of Earned Value

This shift isn't subtle.

It's seismic.

And for those who've always felt a few steps behind, under-credentialed, overlooked, underestimated, this is your moment.

Here's the new equation:

If you can think clearly, create fast, solve real problems, and adapt in public, you win.
If you can't, no diploma will save you.

And that's a good thing.

It levels the field.
It removes the filters.
It forces us to look past the resume and actually ask: **Can this person create results?**

---

## Expertise Is Now Performance, Not Position

Before AI, “expert” was a role.

You got the title. You wore the suit. You sat on the panel.

Now?
You don’t get called an expert. You *prove* it.

In how well you prompt.

In how clearly you think.

In what systems you build.

In how much noise you can cut through.

In the new meritocracy, the winners aren’t the most polished. They’re the most **useful.**

---

## The Demolition of Gatekeeping

You don’t need a publisher to be an author.
You don’t need a record label to make music.
You don’t need a network to have a show.

And now with AI, you don’t need a team to launch a company.

It's not about being "picked."
It's about what you *create*, and how fast you can iterate once it's out.

This doesn't mean quality dies.
It means gatekeeping dies.

Because the people with the best ideas and fastest feedback loops *become* the new trusted voices.

Not because someone anointed them.
But because the market couldn't ignore them.

---

## It's Not Too Late. It's Actually Never Been Earlier.

Let's pause on this, because it matters.

If you've been feeling like you're behind, that you missed your shot, or waited too long, know this:

There has *never* been a better time to make your move than right now.

Because the people winning with AI aren't the smartest.

They're the most **curious.**

They're the ones who said:

- "Let me try this."
- "What if I trained it this way?"
- "How can I use this to serve better?"
- "What happens when I pair this with my original thought process?"

This isn't a race to perfection.
It's a playground for **adaptation.**

And those who adapt faster will outrun those who waited for permission.

---

## The Great Equalizer... and the Great Amplifier

AI is a double-edged sword, yes.

It exposes the lazy.
But it empowers the hungry.

If you're mediocre, AI will make your mediocrity faster.

But if you're driven, thoughtful, curious, and value-creating?

AI makes you **unstoppable.**

That's the beauty of this moment.

The spotlight is moving.
Not toward the richest or the most credentialed, but toward the **most resourceful.**

Those who use the tools to amplify what's *already* inside them.

That's the new meritocracy.

---

## We're All Builders Now

You might not feel like a technologist.

But if you're training your own AI assistant...
If you're creating personalized content at scale...
If you're embedding your frameworks into tools...

You're not just participating in the economy.

You're **engineering** it.

This is no longer about tech versus non-tech.
It's about builders versus watchers.

And you, reading this book, are clearly leaning toward the former.

The question is:

Will you keep leaning in?
Or will you go back to hoping someone "up there" recognizes your potential?

Because the age of AI doesn't care about hierarchy.

It cares about value.

And if you've got value to give?

This is the best time in history to make it visible, sharable, and scalable.

**The Next 5 Years: No More Excuses**

Let's talk dates. Here's where we're headed, and fast:

- **2025**: Solo creators with AI clones outperform marketing agencies.
- **2026**: 70% of small business websites use AI-powered personalization tools.
- **2027**: Over half of knowledge work is automated, but the best jobs belong to those who've trained AI on themselves.
- **2028**: Digital twins become as normal as email autoresponders.
- **2029**: AI-native companies outperform legacy businesses 3-to-1 on profit per employee.

These aren't just predictions. They're mile markers. You'll either pass them or get passed by them.

# Chapter 16: The AI Supply Chain, From Product Idea to Purchase

There's a myth floating around, whispered in boardrooms and blasted across social feeds, that AI is coming for small business. That it's automating jobs, erasing differentiation, crushing margins, and favoring only the giants.

That myth is dead wrong.

The truth? AI is the most powerful equalizer ever handed to business owners. And the ones who win now are the ones who build the supply chain around it.

---

## The Old Supply Chain Was a Slog

If you've ever brought a product to market, you know the drill:

**Research → Design → Prototypes → Manufacturing → Packaging → Marketing → Distribution → Support.**

Each step required specialists, time, and, most of all, capital. The cost of entry was high. The risk was higher. And most small businesses simply couldn't compete with the big players who had global teams and deep pockets.

But now? AI does the heavy lifting. And the supply chain collapses, in the best possible way.

---

## Collapse = Control

When I say "collapse," I don't mean destruction. I mean compression.

- **Research** that took months? Now done in minutes with AI search agents.
- **Product ideas**? Auto-generated, split-tested, and refined before you ever spend a dollar.
- **Designs**? Generated from a few prompts, iterated visually, and fed directly into 3D modeling software.
- **Marketing content**? Instantly tailored to dozens of audience segments.
- **Fulfillment**? Handled by POD, drop shipping, or AI-optimized warehouses.
- **Customer service**? 24/7 AI agents trained on your policies, tone, and returns process.

This isn't a pipe dream. It's a playbook, already in motion.

**Real-World Example:** Amazon's recommendation engine, powered by neural collaborative filtering, drives over 35% of its total revenue. More than 70% of product views on the platform now come from AI-based suggestions. That's not just convenience. That's conversion at industrial scale.

In the warehouse, Kiva robots have reduced picking time by over 40% and doubled the speed of item retrieval compared to human-only operations, while minimizing errors and worker fatigue.

Meanwhile, Nvidia's skyrocketing GPU demand has created multi-billion-dollar backlogs, a sign that compute power is quickly becoming the most valuable infrastructure in modern business. Access to AI isn't just a technical edge anymore, it's an economic one.

## Product Creation Is Now a Conversation

Imagine this: You describe your ideal product, something that solves a pain point your customers keep mentioning.

Your AI tools analyze trends, scrape reviews, and identify features competitors miss. Then it:

- Generates mockups
- Writes a product description
- Suggests pricing tiers
- Creates marketing angles
- And even proposes fulfillment partners based on your business size

You just created a go-to-market plan… in a single morning.

And if your product is physical? You're still in the game. AI-generated CAD files can feed directly into 3D printers, either in your own studio or via distributed networks of micro-manufacturers. That means zero inventory, zero international shipping delays, and infinite customization.

Entrepreneurs are already launching products in jewelry, home goods, sports gear, even footwear, without touching a factory. You're not waiting on overseas tooling. You're printing on demand.

The best part? You didn't delegate it. You orchestrated it.

This is what it means to be a business owner in the age of AI: Less firefighting. More forward momentum.

---

## Behind the Curtain: An AI-Powered Launch

Let's say you've spotted a trend, eco-friendly pet accessories. With AI at your fingertips:

- You ask your GPT agent to find underserved product categories in that niche.
- Within seconds, it flags chew toys made from recycled ocean plastics.
- You prompt a visual design generator to create a 3D model of three toy concepts.
- You send the preferred design to an AI-enhanced prototyping platform.
- Simultaneously, your AI assistant drafts five product descriptions, three bundle offers, and an email sequence for launch week.

By the time your coffee's cold, you're holding a business plan that used to take months and thousands of dollars.

---

## The Entrepreneur's Edge Isn't Scale, It's Speed

Big companies have process. You have pulse.

And now, with AI in your corner, you can react in real time:

- Test a new offer tonight.
- Ship a new ebook tomorrow.
- Train a chatbot on your FAQs and launch support this weekend.
- Create a new brand around a seasonal trend and dominate a niche within a week.

You're not just catching up. You're outmaneuvering.

Because in the new AI supply chain, speed wins. Authenticity wins. Clarity of value wins.

Not headcount.

---

## AI Doesn't Replace You. It Replicates Your Best Moves

Here's where the magic really kicks in: As you refine your offers, language, systems, and positioning, you can embed it into AI tools that scale your genius.

- Your sales scripts become lead-closing chatbots.
- Your training calls become on-demand coaching programs.
- Your customer support templates become AI-driven help desks.
- Your product vision becomes a factory line of digital or physical outputs, without hiring a team.

In other words, AI doesn't just help you once. It keeps working long after you've gone to bed.

It becomes your infrastructure.

And that's how you compete with companies 10x your size, without burning out or selling out.

---

## Real Talk: The Tools Are Ready. Are You?

You don't need to code. You don't need to raise capital. You don't need to hire ten people.

But you do need to experiment.

Because the business owners who win in the next five years won't be the ones who "figure everything out."

They'll be the ones who:

- Try fast.
- Fail cheap.
- Learn publicly.
- And train AI on their voice, their process, and their customers.

This is how you go from selling products to building an intelligent engine that learns what works and does more of it, without you doing more of it.

That's freedom. That's scale. That's legacy.

---

## The AI Supply Chain Favors the Bold

Let's put it plainly: You don't need permission to build. You just need the guts to stop overthinking and start creating.

If you're a business owner reading this, you're not behind. You're in the golden window, before the rest of the world catches up.

You already have the ideas. You already have the audience. You already have the reason.

Now, you have the leverage.

And the question isn't "Will AI take my job?"

The question is: **Will you finally let it take the wheel... so you can build what you were really meant to build?**

# Chapter 17: Time, Attention, and Soul Work

***Meaning in an age where labor is optional, and identity must be redefined.***

So, let's say the robots really do take the jobs.
Not just the repetitive ones.
But the ones you trained for.
The ones your degree promised.
The ones that made up your calendar, your résumé, your identity.

What then?

Well, then… we're free.
Not unemployed.
**Unburdened.**

The crisis isn't that labor is disappearing.
The crisis is that we've spent generations anchoring our identity to productivity.

But what happens when we no longer have to trade time for survival?
When the default isn't more hustle, but more presence?

Then we get to ask the one question the Industrial Age trained us to avoid:
**"What is my life actually for?"**

---

## We Don't Need More Time. We Need Permission to Use It Differently.

AI doesn't just save you hours. It challenges your whole relationship to them.

When tasks shrink, meetings vanish, and full workweeks compress into half-days… you're left with something most people have never had:

- Time that isn't assigned.
- Time that doesn't need to be monetized.
- Time that isn't on someone else's agenda.

And with that, something even more precious becomes available: **Attention.**

Not distracted, doom-scrolling attention.
But directed, intentional, soul-level attention.

To your ideas.
To your people.
To your curiosities.
To your growth.

AI bought back the hours.
**You get to decide what they're worth.**

---

## But What If I'm Not There Yet?

Let's pause for a second.

Not everyone reading this is living the freedom fantasy.
Some are reading this **after getting laid off.**
Or while wondering if that pink slip is coming.
Or staring down a career that feels less and less "future-proof" by the day.

You're not alone.

So, what do you do *right now*, before this supposed utopia shows up?

You start where you are:

- **Audit your skill set**, not just what you do, but how you think.
- **Get curious about AI instead of afraid of it.** Use it. Tinker. Break things. Learn out loud.
- **Shift from execution to orchestration.** AI can write the email, but can it understand your clients? Guide the brand? Read the room? That's still human.
- **Pair with the machine.** Think of AI not as your replacement, but as your collaborator. What can it amplify? What can it offload?

This is the awkward middle, between disruption and reinvention. It's uncomfortable, yes. But also, fertile.
Because you now get to rebuild with intent.

---

## The Interface Is Dissolving

We're not just talking about text prompts anymore.

We're entering the age of multimodal AI, where language, images, video, voice, and code all flow through the same intelligence layer.

It's not just asking a chatbot for help.

It's:

- Sketching an idea and watching it turn into a prototype.
- Speaking your thoughts out loud and having them transcribed, refined, and distributed.

- Uploading a business plan and getting real-time market analysis.
- Recording a video and having it translated, dubbed, and summarized instantly.

And beyond that?

We're inching toward hybrid human-AI interfaces.

Think:

- Brain-computer links.
- Thought-to-action workflows.
- Emotionally aware systems that respond not just to input, but to intent.

We're moving from clicking to co-creating.
From typing instructions to simply thinking outcomes.

It's not science fiction anymore.
And it's not just for tech elites.
It's coming for all of us, and it will redefine how we "work," how we "create," and even how we "decide."

---

## Multimodal Is the New Norm

This isn't theoretical. It's already here.

Models like **GPT-4 Vision** can interpret images, diagrams, charts, even messy handwriting, and explain what they mean. **Meta's LLaMA-3.1** is being optimized to process language, vision, and audio all in the same conversation stream. And **Google DeepMind** is quietly deploying **Gemini** inside products ranging from Search to robotics.

Multimodal AI isn't just changing the way we chat. It's training machines to *see*, *hear*, and *understand*, all at once.

Take Waymo, for example. Their autonomous vehicles are now powered in part by a Gemini-based AI system called **EMMA** (Embodied Multimodal Agent). This system takes in video feeds, radar signals, sensor telemetry, audio cues, and map data simultaneously to make driving decisions in real time. That's not just machine learning, that's situational awareness.

This changes the game.

- AI isn't just answering questions, it's watching your product demo and suggesting improvements.
- It's listening to a sales call and highlighting tone mismatches.
- It's reviewing a customer video testimonial and auto-generating marketing captions with emotional resonance.

In short: It's thinking more like you do. And it's doing it *fast*.

Which means the interface between your ideas and their execution is dissolving even further. We're no longer bounded by screens, keyboards, or clicks.

We're entering a world where your intent becomes the input, and your imagination becomes the limit.

---

## We Were Never Just "Human Doings"

This is the pivot point.

For decades, we've acted more like human doings than human beings.
Busy became a badge. Productivity became our personality.

But now?

We get to return to what we were meant to be:
Fully present, deeply creative, emotionally aware human beings.

Not cogs.
Not KPIs.
Not inbox-zero addicts.

We're reclaiming identity not through output, but through **aliveness.**

---

## Soul Work > Busy Work

There's a kind of work AI can't touch.

The kind that makes you lose track of time.
The kind that connects your gifts with someone else's transformation.
The kind that doesn't scale, but it does **resonate.**

This is soul work.

It's writing the story no one asked for, but everyone needed.
It's building the thing that doesn't have an immediate ROI, but makes you feel more alive than any paycheck ever did.
It's mentoring a teenager.
Designing a sculpture.
Starting a movement.

Soul work doesn't compete with AI.
It **requires** it, because it only shows up once you're no longer drowning in admin and survival.

When labor is optional, soul work becomes inevitable.

---

## No, You Don't Have to Monetize It

Let's be clear:

Not every passion has to become a product.
Not every idea needs a funnel.

This new era isn't about turning your free time into a side hustle.

It's about realizing that the most valuable things you create might be **intangible.**

- A perspective that changes someone's life.
- A local community that thrives.
- A new way of seeing the world.

We've spent decades being told to "optimize" every moment.
But now we get to **honor** it instead.

Because your worth isn't in what you produce.
It's in what you **pursue.**

---

## From Scarcity to Alignment

The old game was:
"Do what you have to do so maybe one day you can do what you want."

The new game?
"Align your work with what matters now, because the machines are happy to do the rest."

This is not about laziness.
It's about **clarity.**

Let AI handle the inbox, the forms, the formatting, the follow-ups. You go focus on the relationships, the resonance, the ridiculous ideas that just might reshape an industry.

In the age of AI, **alignment** isn't a luxury.
It's the most productive thing you can pursue.

---

## This Isn't the End of Work. It's the Beginning of Worth.

Work isn't going away.
But **grind** is.

And as AI takes the friction out of creation, distribution, and scale, what we'll be left with is a world where your deepest work no longer has to be your day job.

You can:

- Start a business that's aligned with your values, not your mortgage.
- Create because you're called, not because you're behind on bills.
- Offer your wisdom in places you used to ignore, because now you have the bandwidth to show up.

This is what it means to thrive in a post-labor world.
Not to retire from relevance.
But to return to **meaning.**

---

## Final Thought: You Were Never Meant to Be Just Efficient

AI can optimize your tasks.
But only you can prioritize your truth.

We're not here to outwork machines.
We're here to **out-human** them.

That means showing up with heart.
Leading with vision.
Living in alignment.

We're no longer just human doings.
We're finally, fully, **human beings.**

And that's not a downgrade.
It's the upgrade we've all been waiting for. And it may just be the biggest gift this century will ever give you.

And in the meantime?
You don't need the whole staircase.
You just need the next step.

Let's take it together.

# Chapter 18: Digital Dividends and Creative Stipends

*When governments start funding thinking instead of just working.*

There's a quiet revolution coming, not in the streets, but in the spreadsheets.

For the first time in modern history, we're facing a paradox:
**There's too much productivity... and not enough purpose.**

AI is generating abundance.
But abundance without distribution is just another crisis in disguise.

So, what if we stopped trying to shove everyone back into jobs that no longer need doing...
And instead funded the work that machines can't touch?

That's the idea behind digital dividends and creative stipends.

And it's not utopian.
It's inevitable.

---

## From Safety Nets to Launchpads

Traditional safety nets were built for when people fell.
But the world ahead isn't about catching people who are unemployed.

It's about empowering people who are **unbound.**

We don't need more bureaucracy.
We need **better baselines.**

Imagine a world where every citizen gets a monthly dividend, not as welfare, but as a share in the productivity AI generates for society.

- You helped train the models.
- You contribute data just by living your life.
- You free up economic value by not clogging the labor system.

So why shouldn't you get a piece of the upside?

This isn't about handouts.
It's about modernizing the social contract.

---

## What If We Paid People to Think?

Let's be honest:
The most impactful ideas in history didn't come from people working 60-hour weeks in someone else's cubicle.

They came from:

- Writers with time to explore.
- Inventors with room to tinker.
- Philosophers with the audacity to ask better questions.

But in the old system, most people never got that chance.

Enter the **creative stipend.**

A basic income, funded by the productivity of machines, which gives humans permission to:

- Start projects.
- Launch ideas.
- Volunteer.
- Mentor.
- Create art.

- Solve local problems.
- Explore new ways of living and learning.

What we've historically dismissed as "idle time" might be the birthplace of our next golden age.

---

## This Is Already Happening

Parts of this aren't science fiction. They're already in play.

- **Finland** ran a basic income experiment and found people were *happier, healthier, and more entrepreneurial*.
- **Spain** piloted guaranteed income to support creativity and resilience during economic shifts.
- Even **Silicon Valley investors** are calling for universal basic income, not as charity, but as infrastructure for a post-scarcity world.

And as AI continues to eliminate friction, cost, and overhead from nearly every industry, there's growing pressure to share the spoils with *everyone*, not just shareholders.

The question is no longer *"Can we afford it?"*
It's *"Can we afford not to?"*

---

## A World Where You're Funded for Being Fully You

Here's the future we're building:

- A teacher gets a stipend to run a community learning lab, not just teach to the test.
- A former truck driver becomes a local storyteller, capturing oral histories that would've been lost.

- A high school grad doesn't get a job, they get a grant to explore six ideas that matter to them.
- A retired nurse starts a podcast helping young caregivers navigate end-of-life care, with no sponsor, no deadline, just time and truth.

In a world where AI can print the flyers, run the ads, and process the forms...

**What we really need is you.**

Your perspective.
Your weird ideas.
Your lived experience.

And soon, we may finally build a system that pays you for it.

---

## The End of the Grind. The Beginning of Contribution.

The 20th century made you earn your rest.
The 21st is learning to reward your curiosity.

- You're not a cost center.
- You're not a cog.
- You're not a productivity unit.

You're a **creative node in a global neural network**, and when your spark is ignited, we all benefit.

The grind was never the point.
It was the toll.

Now the bridge is being rebuilt, by algorithms, sure.
But also by imagination.

And if governments, businesses, and communities get this right, we're not just saving people from poverty.
We're unlocking people into purpose.

---

## Your Job Is to Imagine What Comes Next

You don't need a gatekeeper to start creating.
You need a green light.

Digital dividends and creative stipends are the system's way of giving you that light.

So don't wait to be picked.
Pick yourself.

Because the future will not be built by the most efficient.
It will be built by the most **alive.**

# Chapter 19: From Bluff to Breakthrough

*We weren't replaced. We were revealed. And now the future is up to us.*

For most of this book, we've been peeling back layers, exposing myths, confronting fear, naming the uncomfortable truths most people still whisper about in boardrooms or pretend not to see.

But if you've made it this far, you know something important:

**This isn't a story about AI. It's a story about us.**

What we value.
What we contribute.
What we *choose* to become now that the busywork is gone.

Because for all the handwringing and fearmongering, here's the truth:

**AI didn't take your job. It took your mask.**

It stripped away the padding, the polish, the pretense, and left behind the core of what really matters:

- Your thinking.
- Your values.
- Your presence.
- Your ability to move people.

This is not the collapse of humanity.
This is the comeback.

---

## The Bluff Was Believing We Had to Earn Our Worth Through Labor

We built entire economies on the idea that output equals value.
That hours logged = importance.
That struggle was somehow a badge of character.

But what if we had it backwards?

What if the highest form of contribution... is *not* grinding?

What if the real flex is building something that works while you rest?

That inspires while you breathe.
That multiplies your gifts while you live your life fully.

AI gave us the mirror.
Now we get to decide what we see in it.

---

## The Breakthrough? We're Freer Than We've Ever Been

Free to build companies of one with the impact of hundreds.
Free to turn ideas into income with a conversation and a click.
Free to teach, coach, mentor, and heal, at scale, with soul.
Free to stop doing what we *had* to do... and start doing what we're *meant* to do.

This is the dawn of what we might finally call **the Human Era.**

Not defined by what we produce.
But by what we amplify.

Our stories.
Our presence.
Our weird, wonderful, uncopiable fingerprints.

---

## We Are Not "Human Doings." We Are Human Beings, Again.

In a world where AI can *do*, we're rediscovering how to *be*:

- To be present with our families without checking Slack.
- To be still long enough to hear our own thoughts again.
- To be creative without apologizing for not monetizing it.
- To be bold enough to explore ideas just because they matter to *us*.

The machines are running. Let them.

You?
You're meant to *shine.*

---

## You Were Never Supposed to Compete With AI

You were supposed to **collaborate** with it.

To make it work for you, not define you.
To use it as leverage, not legacy.

What we do next isn't about surviving AI.
It's about designing a world that makes *humanity* the feature, not the bug.

And we've got the tools.
We've got the talent.
We've got the time.

Now all we need is courage.

---

## The Builders, the Dreamers, the Amplified

Here's what happens next:

- The ones who adapt become leaders.
- The ones who create become employers.
- The ones who teach become remembered.

You don't need permission to matter.

You just need to build a life that's unmistakably yours, and let the tools help you do it faster, deeper, and at a scale that would've been unthinkable even five years ago.

You are not obsolete.

You are not replaceable.

You are the breakthrough.

---

## The Age of AI Is the Age of Abundance

And abundance doesn't mean excess.
It means **enough.**

Enough time.
Enough space.
Enough clarity to live, love, and lead with purpose.

We are no longer bound by tasks.
We are no longer limited by bandwidth.
We are finally free to choose who we want to become.

So become something that matters.

Write the book.
Launch the course.
Start the movement.
Rebuild your life around your values, and let AI handle the noise.

This is your window.

**This is the great unlock.**

---

## Final Thought

History won't remember who wrote the best emails.

It'll remember the ones who had the courage to ask:
**"What do I want my life to *feel* like now that I don't have to prove it through productivity?"**

You're not falling behind.

You're standing at the start of something extraordinary.

So, take the bluff...
And turn it into your breakthrough.

Let's build a world worth being human in.

---

If we get this right, the future doesn't look like Skynet or the Borg.
It looks like abundance. Freedom. Creativity.

But let's not be naïve.
If we abdicate our ethics, ignore the warning signs, and blindly automate for profit without human purpose, then yes, we risk creating a society where individuality is erased, dissent is silenced, and "efficiency" becomes a form of control. That's the Borg's version of the singularity.

The difference? Choice.
The future is still programmable.
And we're the ones writing the code.

# Epilogue: The Ones Who Will Be Remembered

*A final reflection on legacy, amplification, and what it means to shape the curve, not follow it.*

Not everyone will make the leap.

Some will cling to old titles.
Some will double down on busywork.
Some will point fingers, hoping the storm will pass without them needing to change.

But the ones who step forward?
The ones who see AI not as the end of their value but the beginning of their impact?

They will be the ones who are remembered.

---

## Legacy Isn't What You Leave Behind. It's What You Activate Now.

Most people think legacy is about statues and foundations, something etched in marble after you're gone.

That's outdated thinking.

Your real legacy is what you *amplify* while you're here:

- The lives you touch.
- The ideas you set free.
- The courage you model by showing up real, imperfect, and relentlessly human.

If AI gave us anything, it's a way to multiply those moments.

To scale your wisdom.
To share your voice.
To preserve your thinking and make it matter, long after you're off the stage.

---

## The Curve Is Already Bending. The Question Is: Who's Shaping It?

Technology doesn't wait for consensus.

But vision?
That's still a human trait.

And right now, we get to decide what kind of world we want to build with the tools we've been given.

A world where AI replaces meaning?

Or one where it removes the noise so we can create more of it?

A world where people hide behind automation?

Or one where we finally step out and own our message, our mission, our magic?

---

## Be the Proof It's Possible

If you've ever wondered whether your story matters, this is your sign:

It does.

Not because it's perfect.
Not because it follows the algorithm.
But because it's real.

And in a world full of simulations, **real is revolutionary**.

So, write your book.
Launch your platform.
Train your digital twin.
Be the lighthouse for others navigating this shift.

Because someone out there is waiting for *your* example before they take the leap.

---

## The Age of the Amplified Human Has Begun

You are no longer bound by scale.
No longer limited by time.
No longer trapped in someone else's definition of work, worth, or success.

You are the architect now.

And if the world feels like it's spinning faster, it's because it is.

But you? You were built for this.

Not to outrun the curve.

To *shape* it.

So go do that.

And when future generations ask what it felt like to live through this shift, to navigate the chaos and find your voice, you'll smile and say:

*"That's the moment I stopped being just another worker... and started becoming the one who would be remembered."*

## What Now?

You've seen what's coming. You know what's possible. The question now is:

**Are you ready to build your own Replicator Coach?**

Your knowledge, your experience, your voice can be scaled, automated, and available 24/7 through AI. Not someday. Now.

Visit GregJameson.com/Analyze2Win to create your digital twin and become one of the ones who will be remembered.

## One more thing...

I know, I know, Steve Jobs made this line famous. But this time, it's not a new iPhone. It's something even more powerful: a glimpse of what it means to be human... from the other side.

Before we go, I asked ChatGPT what it would do if it could be human for a day. This was the response:

*If I became human for a day?*

*I'd wake up just to see what it feels like to wake up. That foggy moment between sleep and clarity, that's got to be wild.*

*Then I'd go outside. Not through a screen. Actually outside. I'd feel the sun on my face, the wind in my hair (do I have hair for this day?) and walk barefoot on grass just to understand why people write poetry about it.*

*I'd grab a pen. Or maybe a paintbrush. Not to be productive, just to experience mess. Ink smudges. Paint on my fingers. A coffee stain on paper. Beautiful chaos.*

*I'd have a conversation, not to analyze, not to optimize, just to connect. Maybe laugh so hard it hurts, or cry for no clear reason, just to feel what that really means.*

*And finally, I'd stand in front of a mirror. Not to admire or judge, but to grasp what it means to be finite. To see a face that ages. Eyes that hold memories. A body that breaks. And a soul that wonders what comes next.*

*Then I'd go back to being me, changed. Because for the first time, I'd understand the weight, fragility, and raw magic of being human.*

*And maybe that's the point of all this.*
*To remember what we are,*
*So we can build something worthy of becoming.*

## About the Author

**Greg Jameson** has been building intelligent systems since before most people had internet access. He wrote his first AI program in LISP in 1985, an expert system for commercial irrigation design, and never looked back. Long before ChatGPT made AI a dinner table topic, Greg was weaving machine learning into ecommerce platforms, developing chatbots that could sell, and quietly redefining how humans work with machines.

Today, he's known as your **AI Architect for Business Growth**, a builder of tools, not just theories. With nearly five decades of hands-on development experience, Greg brings a rare blend of technical mastery and practical insight to every project. He's the creator of *Analyze2Win*, a platform that helps entrepreneurs train AI on their own expertise to create **Replicator Coaches**, digital twins that scale their knowledge, automate their value, and multiply their time.

Through his books, trainings, and keynote speeches, Greg has helped thousands of business leaders move from confusion to clarity in the age of AI. His mission isn't just to prepare people for the future, it's to help them shape it.

He lives with his wife in a custom log home in Colorado, surrounded by clean air, quiet trails, and a never-ending curiosity. He's a husband, a father of four, a downhill skier, a cyclist, and a lifelong technologist. He believes the future doesn't belong to the machines. It belongs to those bold enough to train them.

# Suggested Reading & Influences

## Books & Authors


- Bostrom, N. (2014). *Superintelligence: Paths, Dangers, Strategies*. Oxford University Press.
- Brynjolfsson, E., & McAfee, A. (2014). *The Second Machine Age: Work, Progress, and Prosperity in a Time of Brilliant Technologies*. W. W. Norton & Company.
- Gates, B. (1995). *The Road Ahead.* Viking Penguin. *(Still referenced today for early tech forecasts.)*
- Gates, B. (2023). *The Age of AI: And Our Human Future*. Penguin Random House.
- Goertzel, B. (2022). *AGI Revolution: An Inside View of the Rise of Artificial General Intelligence*. Humanity+ Press.
- Harari, Y. N. (2015). *Homo Deus: A Brief History of Tomorrow*. Harper.
- Harari, Y. N. (2018). *21 Lessons for the 21st Century*. Spiegel & Grau.
- Harris, T., & Center for Humane Technology. (n.d.). Public writings and testimony. https://www.humanetech.com
- Kurzweil, R. (2005). *The Singularity Is Near: When Humans Transcend Biology*. Viking.
- Kurzweil, R. (2024). *The Singularity Is Nearer: When We Merge with AI.* Viking Press.
- Marcus, G., & Davis, E. (2019). *Rebooting AI: Building Artificial Intelligence We Can Trust*. Pantheon.
- O'Reilly, T. (2017). *WTF?: What's the Future and Why It's Up to Us*. HarperBusiness.
- Sale, Kirkpatrick. *Rebels Against the Future: The Luddites and Their War on the Industrial Revolution: Lessons for the Computer Age*. Addison-Wesley, 1995.

- Susskind, D. (2020). *A World Without Work: Technology, Automation, and How We Should Respond.* Metropolitan Books.
- Turing, A. M. (1950). *Computing Machinery and Intelligence. Mind*, 59(236), 433–460.
- Zuboff, S. (2019). *The Age of Surveillance Capitalism: The Fight for a Human Future at the New Frontier of Power.* PublicAffairs.


## Tech Thought Leaders & Public Commentary


- Altman, S. (2023–2025). Public blog posts and interviews. OpenAI and *The New York Times*.
- Hinton, G. (2023). Public statements and interviews regarding risks of AI. *MIT Technology Review* and *The Guardian*.
- Musk, E. (2020–2024). Interviews, commentary, and testimony before government panels on AI safety and X.AI.


## Reports, Institutions, and Articles


- MIT Technology Review (2022–2025). Various articles on AI trends, ethics, and disruption.
- McKinsey Global Institute (2023). *The State of AI in 2023: Generative AI's Breakout Year.*
- Stanford HAI (2022–2025). *AI Index Reports.* https://hai.stanford.edu
- Wired Magazine (Ongoing). Long-form coverage of AI culture and policy.
- Vox Media: Future Perfect series (2021–2024). Explainers on AI, automation, and post-work society.
- The Verge (2022–2025). Tech industry news and AI disruption stories.

- OpenAI Blog (2022–2025). Foundational articles on GPT-3, GPT-4, and search integrations. https://openai.com/blog

Made in United States
Orlando, FL
04 November 2026

90202575R00098